Imperiled Waters, Impoverished Future: The Decline of Freshwater Ecosystems

JANET N. ABRAMOVITZ

Anjali Acharya, *Staff Researcher*

Jane A. Peterson, *Editor*

WORLDWATCH PAPER 128
March 1996

FINANCIAL SUPPORT is provided by Carolyn Foundation, the Nathan Cummings Foundation, the Geraldine R. Dodge Foundation, the Energy Foundation, The Ford Foundation, the George Gund Foundation, The William and Flora Hewlett Foundation, W. Alton Jones Foundation, John D. and Catherine T. MacArthur Foundation, Andrew W. Mellon Foundation, The Curtis and Edith Munson Foundation, Edward John Noble Foundation, The Pew Charitable Trusts, Lynn R. and Karl E. Prickett Fund, Rockefeller Brothers Fund, Rockefeller Financial Services, Surdna Foundation, Turner Foundation, U.N. Population Fund, Wallace Genetic Foundation, Weeden Foundation, and the Winslow Foundation.

PUBLICATIONS of the Institute include the annual *State of the World*, which is now published in 27 languages; *Vital Signs*, an annual compendium of global trends that are shaping our future; the *Environmental Alert* book series; *World Watch* magazine; and the Worldwatch Papers. For more information on Worldwatch publications, write: Worldwatch Institute, 1776 Massachusetts Ave., NW, Washington, DC 20036; or fax 202-296-7365; or see back pages.

THE WORLDWATCH PAPERS provide in-depth, quantitative and qualitative analysis of the major issues affecting prospects for a sustainable society. The Papers are written by members of the Worldwatch Institute research staff and reviewed by experts in the field. Published in five languages, they have been used as concise and authoritative references by governments, nongovernmental organizations, and educational institutions worldwide. For a partial list of available Papers, see back pages.

DATA from all graphs and tables contained in this Paper are available on 3 1/2" Macintosh or IBM-compatible computer disks. The disks also include data from the *State of the World* series, *Vital Signs*, *Environmental Alert* book series, Worldwatch Papers, and *World Watch* magazine. Each yearly subscription includes a mid-year update, and *Vital Signs* and *State of the World* as they are published. The disk is formatted for Lotus 1-2-3, and can be used with Quattro Pro, Excel, SuperCalc, and many other spreadsheets. To order, see back pages.

Table of Contents

Sections of this paper may be reproduced in magazines and newspapers with written permission from the Worldwatch Institute. For information, call the Director of Communication at (202)452-1999 or Fax: (202)296-7365.

The views expressed are those of the author and do not necessarily represent those of the Worldwatch Institute, its directors, officers, or staff, or of its funding organizations.

ACKNOWLEDGMENTS: I would like to express my appreciation to a number of people for their contributions. Thanks to the following reviewers for their thoughtful comments on various drafts of this paper: Dirk Bryant, Chris Frissell, Les Kaufman, Melannie Stiassny, David Wilcove, Robin Welcomme, and John Young. Thanks also to the following individuals and organizations for providing data: Kay Brown, Dave Harrelson, John Harrison, James Lichatovich, Lars Mobrand, Paul Parmalee, Paul Summers, Tania Williams of the National Research Council, and Larry Masters of The Nature Conservancy. Jane Peterson deserves special thanks for editing. Many thanks to colleagues at Worldwatch: Ed Ayres, Chris Bright, and Chris Flavin for review and collegial support; Tonje Vetleseter for research assistance; the library staff, Lori Ann Baldwin and Laura Malinowski, for tracking down the often obscure; Jennifer Seher for layout and design; and Jim Perry, Denise Byers Thomma, and Tara Patterson of the communication department for coordinating production and outreach. Finally, thanks to all those who enthusiastically encouraged the idea of addressing freshwater biodiversity.

JANET N. ABRAMOVITZ is a Senior Researcher at the Worldwatch Institute where she focuses on biodiversity, natural resources management, human development, and social equity. She is a co-author of the Institute's annual reports, *State of the World 1996* and *Vital Signs,* and writes regularly for its magazine, *World Watch.* Prior to joining Worldwatch she worked at World Resources Institute and other governmental and nongovernmental organizations. She has published extensively on biodiversity and gender issues.

Introduction: The Biodiversity Deficit

By the time Europeans arrived in the New World, the streams, rivers, and lakes of Eastern North America were home to over 300 species of mussels—fully one-third of the world's freshwater mussel species. During their 30- to 130-year life span, these sedentary creatures remain in the same spot, filtering vast quantities of water to glean microscopic plankton. Their cleansing ability helps provide the kind of water quality that other species, including the fish on which we rely for food, require in order to prosper. The diversity and abundance of mussel species enabled them to live in balance with their predators for millennia and to rebound from periodic natural disasters.[1]

Mussels have an extraordinarily intricate life cycle. Each mussel species depends on a particular fish species during a brief but critical part of its life. Larvae released by the female mussel must attach to the gills of the host fish for a short period. After further development, the larvae then drop off and, if they land in a hospitable habitat, develop into juvenile mussels. But, if the host fish disappears, the mussels cannot reproduce. And if the mussels die, the water will not be adequately cleansed and the entire ecosystem may collapse.[2]

For thousands of years North America's aboriginal peoples harvested huge numbers of mussels without any apparent damage to the species. Yet, in this century, human beings have unwittingly brought many mussel species to the verge of extinction. Since 1900, about 10 percent of the mussels species in North America have become extinct; 67 percent of the remaining 297 species and subspecies are endangered, threatened, or otherwise at risk; and some are represented only by an aging population that no longer reproduces. Only 25 percent are considered stable.[3]

Throughout their range, the greatest toll on these sensitive organisms has come from habitat destruction caused by plowing and deforesting the landscape, draining and polluting the waters, and damming and dredging the rivers. By the turn of the century, some species were already in decline as a result of the rivers being used as sewers. Then, beginning in the 1930s, a spate of dam construction precipitated the loss of many species, altered the flow of many rivers, and cut many mussels off from their host species. After a dam on the upper Mississippi River blocked the migration of the skipjack herring, for example, both the ebony shell mussel and the elephant ear mussel had no means for development and dispersal of their young, and so disappeared from the river above the dam. In a pattern repeated many times over, 50 native mussel species once inhabited the Little Tennessee River; after impoundment by a dam in 1979, only 6 remained.[4]

Compounding these problems, carving navigation channels and mining gravel beds have gradually destroyed mussels' "nesting" sites over the past century. Heavy soil erosion caused by agriculture and construction often buries the immobile creatures in layers of silt and "suffocates" them by clogging their gills.

In addition, mussels have at times been overharvested—they were widely used for manufacturing buttons in the decades before plastic was invented—further decimating some species. Today, North American mussel shells are exported to Japan and ground up for "seed" to start cultured pearls. The recent invasion of non-native mussel species such as the Asian clam and the zebra mussel, which are spreading rapidly into North America, could deliver the fatal blow to remaining members of the world's most diverse freshwater mussel fauna.

The ecological history of North American mussels reflects the fate of freshwater ecosystems around the world.

FIGURE 1

Distribution of Fish Species by Realm

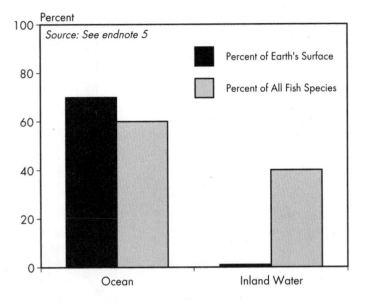

As biological assets, freshwater systems are both disproportionately rich and disproportionately imperiled. Twelve percent of all animal species, including 41 percent of all recognized fish species, live in the 1 percent of the earth's surface
that is fresh water. (See Figure 1.) And yet, at least 20 percent of all freshwater species have become extinct, threatened, or endangered in recent years. In North America, the
continent that has been studied most thoroughly, 67 percent of mussels, 65 percent of crayfish, 37 percent of fish,
and 38 percent of amphibians are either in jeopardy or—in
some cases—already gone. By contrast only 16 percent of
mammals and 14 percent of birds are extinct or at risk.[5] (See
Figure 2.)

Until recently, most of us have had little inkling of the
ecological, economic, and social consequences of altering
watersheds and the natural course of water. We see rivers as
being too vast to be harmed, fish as too plentiful to be per-
manently destroyed. And if certain flora and fauna are

FIGURE 2

Species Extinct and at Risk in North America

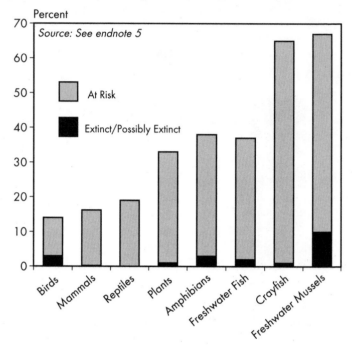

extinguished by building a dam or draining a wetland, so the thinking goes, substitutes will be found—after all, extinction has occurred since life began, and nature has always recovered.

The problem is the sheer scale of the current human assault on freshwater ecosystems. When an ecosystem is destroyed, the life that existed there disappears along with the services once provided. As this scenario repeats itself thousands of times over, the planet is becoming biologically impoverished at every level—genes, species, assemblages, ecosystems, and processes—losing the resilience that has always enabled it to recover from adversity and to create new life. The living library of options for adapting to local and global change is rapidly shrinking.

More alarming than the actual *number* of species that are

now endangered or extinct is the fact that the *rate* at which they are vanishing exceeds natural extinction rates by 100 to 1,000 times. And if now threatened species disappear in the next century, extinction rates could reach 1,000 to 10,000 times prehuman levels. Thus, we are running a "biodiversity deficit," destroying species and ecosystems faster than nature can create new ones. Such a course is even less sustainable than a financial deficit because extinction truly is forever. With less than two million of the tens of millions of species in the world named, much less understood, many species and habitats are being lost before they are even identified. [6]

Nowhere today is the destruction of ecosystems or the loss of biodiversity as acute as in the vast, species-rich tropics, where habitat for freshwater species is being rapidly lost. Tropical fish represent the last big gap in the global catalog of vertebrate diversity. The Amazon, for example, widely known for its rich tropical forests, is also the world's largest river basin and home to a rich array of freshwater fish species, perhaps one-third of the earth's total, many still unnamed. Along the populous Mekong River, new species are still being found and the complex behavior and habitat needs of even the economically important fish species are dimly known.[7]

When we jeopardize a freshwater ecosystem's integrity— its physical, chemical, and biological elements and processes—we compromise its ability to support species and provide the products and services we depend on, services such as controlling floods, purifying water, recharging aquifers, restoring soil fertility, supporting recreation, nurturing fisheries, and supporting evolution. Once nature can no longer provide, we must either do without or try to substitute, usually much less effectively and at much higher cost.

The estimated monetary value of some services provided by intact ecosystems gives an idea of how costly their loss can be. For example, the value of mangrove swamps for flood control and storm protection alone has been figured at $300,000 per kilometer in Malaysia, the cost of rock walls that would be needed to replace them. For example, the worth of a 223,000-hectare swamp in Florida has been calculated at $25 million a year for its services of storing water and recharging the aquifer. And yet, at least half of the original wetlands in the U.S. (excluding those in Alaska) have been drained; the amount lost between the 1780s and the 1980s averaged more than 24 hectares (60 acres) per hour for every hour of those 200 years.[8]

Given human dependence on freshwater products and services, the loss of biodiversity has enormous ramifications. All of our food, regardless of where it is grown, has its origins in the wild. And wild relatives of domesticated crops, livestock, and fisheries will continue to be important to ensuring a sustainable food supply. Even with the recent growth in fish farming, we still depend on nature to provide 86 percent of the aquatic bounty that we consume. Nevertheless, we continue to degrade the foundation that supports the harvest. In the North American Great Lakes, for example, of the 11 species that once provided a commercial catch of 1.4 million kilograms per year, 4 are extinct and the other 7 are at risk. Seeing fish in markets, consumers may be unaware of the magnitude of the changes in freshwater systems.[9]

Freshwater ecosystems are the critical link between land and sea, in effect forming the planet's circulatory system; virtually every human action is eventually reflected in them. Unfortunately, the harmful consequences of our activities on these once-bountiful systems add up and compound each other. Thus, efforts to sustain aquatic ecosystems and ourselves will require a better understanding of their dynamic nature and a shift from today's piecemeal and short-sighted management of natural resources toward managing them in ways that ensure their long-term viability. The prospects for evolution and human well-being are bound up in the health of freshwater ecosystems.

Restructuring the Rivers

The rationale for constraining the free-flowing nature of rivers stems partly from the widespread beliefs that floods serve no beneficial purposes and that undeveloped rivers and their floodplains, wetlands, and backwaters are wasted and unproductive. Such views could not be further from the truth. A dynamic equilibrium exists between the biological and physical features of aquatic systems. Many river systems are adapted to what scientists call "pulse" disturbances—events such as naturally occurring seasonal floods. In some places they may last a few weeks; in others, most of the year. The flood pulses help maintain the natural interactions between a river and its surrounding landscape that make them both extremely productive and diverse—much more so than either area would be if cut off from the other.[10]

Animals and plants across the landscape are adapted to this regime. For example, many fish use the floodplain as a spawning ground and nursery for their young; some consume and help disburse seeds; while others depend on the temporary abundance of food to sustain them for the entire year. Likewise, many plants take advantage of the flood period to germinate and absorb newly available dissolved nutrients, and migratory waterfowl rely on its bounty as well. Many soils need the regular addition of nutrients and organic matter that the floods bring, while other soils contribute materials to the water. People living in these areas have also adapted to this natural rhythm and have learned how to use its bounty to ensure reliable and sustainable livelihoods. To paraphrase fisheries biologist Peter Bayley, the flood pulse is not a disturbance; flood prevention is.[11]

Considerable fragmentation of rivers has already taken place in industrialized countries. In the northern third of the globe (Europe, the former Soviet Union, and North America north of Mexico), dams, water flow regulation from reservoir operation, irrigation, and interbasin transfers have significantly altered 77 percent of the total water discharge of the 139 largest river systems. The few unaffected rivers are in the furthermost northern areas. All of the large rivers of the United States are fragmented—only 2 percent of the country's 5.1 million kilometers of rivers and streams remain free flowing and undeveloped. But the fragmentation and regulation of rivers is not just a problem in industrialized countries—it is occurring throughout the world. It is predicted that two-thirds of the world's total stream flow will be regulated by the end of the decade.[12]

One of the earliest changes humankind made to the structure and function of rivers was to create waterways for navigation. Like the terrestrial road network that expanded over the centuries from footpaths to dirt roads to today's superhighways, the network of aquatic highways has grown in size and technological sophistication. The length of rivers altered for shipping has increased tremendously, from just under 9,000 kilometers in 1900 to about 500,000 today.[13] (See Table 1.)

As navigation and ports have grown, more people and industries have been drawn to river banks and basins. From ancient times to the present, access to trade routes and commerce has been a driving force in settlement and economic growth. Constructing ports, straightening and deepening channels, and regulating the flow of water help maintain this trade infrastructure.

The impacts of navigation and engineering have increased as well. They invariably cut off the main channel's interaction with surrounding habitat, thus fragmenting ecosystems into nonviable segments. Removing stream debris, stabilizing banks, and eliminating riparian vegetation destroy habitat important to many species. Altering the structure of a river also brings about changes in the water depth, flow rate, temperature, sediment content, chemistry, and oxygen concentration—factors that profoundly influence the composition and abundance of species. Compounding these problems, eliminating natural barriers such as waterfalls and using canals to link previously unconnected water bodies have allowed the invasion of non-

TABLE 1

Alteration and Use of Freshwater Systems Worldwide, 1680–1985

Alteration and Use	1680	1800	1900	1950	1980	1985
Waterways altered for navigation (km)	< 200	3,125	8,750	—	498,000	—
Canals (km)	5,000	8,750	21,250	—	63,125	—
Large dams (>15m in height)						
World	—	—	—	5,270	—	36,562
China	—	—	—	2	—	18,820
Large reservoirs						
Number	—	—	41	539	—	1,777
Volume (km²)	—	—	14	528	—	4,982
Water withdrawal (thousand km³/yr)	104	243	654	1,415	3,640	—
Per capita use (thousand m³/yr/person)	153	254	396	563	824	—

Source: See endnote 13

native species. Both the channels and the ships that ply them serve as conveyor belts for these "exotics," with often catastrophic consequences. A few case studies will illustrate just how destructive aquatic highways can be.

In Europe, a number of signs clearly indicate that the Rhine River has suffered from too much human interference. The Rhine runs 1,320 kilometers from its beginning in the Swiss Alps through France, Germany, and The Netherlands out to the North Sea—through the most heavily populated and industrialized part of Europe. Twenty percent of the world's chemicals are produced there. Because the river and its tributaries provide 20 million people with drinking water, most public attention to the river over the past few decades has focused on reducing water pollution.[14]

Now other signs of more fundamental problems have forced a dramatic change in the way the river is viewed and managed. A major flood in early 1995 precipitated the evacuation of 250,000 people in The Netherlands and cost 1.6 billion deutsche marks ($1 billion) in Germany alone. In fact, flooding over the last 20 years has grown significantly more frequent and severe due to increased urbanization, river engineering, and poor floodplain management. At Karlsruhe, for example, a German town on the French border, prior to 1977 the Rhine rose 7.62 meters above flood level only four times this century. Since then, it has reached that level ten times.[15]

In its natural state, the mature Rhine River and its branches meandered through a broad floodplain of fields and forest. Then, in the mid-1800s, a German engineer began the process of creating a single, deep, well-defined river highway, 100 kilometers shorter, to speed transportation to the sea and to facilitate growth and industrialization. Many more recent changes have created a fully engineered river with huge locks, levees, and dams. Ten hydroelectric power plants have been built in the upper Rhine. Draining marshland and walling off the river from the floodplain with concrete barriers provided land for farming, housing, and industry and created some immediate benefits. But

because water could no longer spread out and slowly pene-
trate into the ground, the floodplain could no longer absorb
water, and groundwater levels dropped.[16]

Ironically, containing a river in embankments, reservoirs, and other structures does not reduce the volume of
flood water. What it does do is dramatically increase the rate
of flow, and thus its force downstream.
Today the Rhine River is cut off from
90 percent of its original floodplain in
the upper Rhine, and flows twice as fast
as before. This altered river is digging a
deeper channel—up to 8 meters deeper
than its natural bed—as it speeds its
way to the Low Countries. Thus it is
no wonder that some say these structures should not be called flood control
but rather "flood threat transfer mechanisms."[17]

Part of the Rhine Action Plan is to restore some of the ecosystem to retrieve the benefits of its natural func- tions.

The physical and chemical changes
in the Rhine River system have also
eliminated most of its fish. The Rhine once supported
vibrant fisheries that fed and provided employment for people all along its length. One hundred years ago, 150,000
salmon were caught in The Netherlands and Germany
alone. By 1920 the catch had fallen below 30,000, and by
1958 it had completely disappeared. The 15 salmon—individual fish, not number of species—found in the Rhine
today are believed to have escaped from a Norwegian aquaculture operation. A major goal of a regional program
launched in 1987—the Rhine Action Plan—is a return of
salmon by 2000.[18]

Part of the plan is to restore some of the river ecosystem
in order to retrieve the benefits of its natural functions, such
as providing safe drinking water, recharging groundwater
supplies, and moderating floodwaters. For example,
Germany and France agreed in 1982 to create flood mead-
ows upstream to reduce flooding downstream. The
Netherlands, much of which is below sea and river level and

absorbs the full power of the Rhine's floods and toxic load, is taking steps to rehabilitate some of its floodplain and delta lands. The ultimate goal there is to return 15 percent of the farmland to functioning floodplain. By reviving some of these nonstructural methods of flood control, the region hopes to reduce the destructive consequences of the regulated Rhine River and restore some of the natural system and the life it supports—and thus to minimize the threat and cost of future floods.[19]

Like the 1995 flooding of the Rhine, the inundation of the upper Mississippi and Missouri rivers in 1993 provided a dramatic and costly lesson on the effects of treating the natural flow of rivers as a pathological condition. The Great Midwest Flood of 1993 was the largest and most destructive in modern U.S. history. It set records for amounts of precipitation, upland runoff, river levels, flood duration, area of flooding, and economic loss. The flood waters rendered ineffective more than a thousand levees spanning nearly 10,000 kilometers. In hindsight, many now realize that the river was simply attempting to reclaim its floodplain. Not surprisingly, 1993 was a record spawning year for native species. The flood's human and economic costs combined with its benefits to the ecosystem's functions inspired a rethinking of the way in which large rivers are managed.[20]

The rising frequency and severity of floods occurring along the Mississippi River and its tributaries in recent years brought home the lesson that massive expenditures on flood control through engineering have actually increased the frequency and severity of floods while crippling the river's ability to support native fauna and flora. Not only the construction of thousands of levees and the creation of deep navigation channels, but also extensive row crop farming in the floodplain, and the draining of more than 6.9 million hectares of wetlands (more than an 85 percent reduction in some states) have cut off the ability of the Mississippi's floodplains to absorb and slowly release rain, flood water, nutrients, and sediments. Such ecosystem mismanagement comes at great economic cost, as demonstrated by the decline of aquatic species, the subsidence of the Mississippi Delta, and the enormous financial toll of the 1993 flood.[21]

Today's problems reflect the cumulative impacts of more than one hundred fifty years of actions by public and private interests to expand agriculture, facilitate navigation, and control flooding. Levees, locks, dams, and reservoirs work to keep the river (and its tributaries such as the Missouri and Illinois rivers) in a course not of its choosing. Nearly half of the 3,782-kilometer-long Mississippi flows through artificial channels. Records show that the 1973, 1982, and 1993 floods were substantially higher than they might have been before structural flood control began in earnest in 1927. Measured in constant dollars, damages from the 1927 flood were estimated at $236 million, while those from the 1973 flood stood at $425 million. Financial costs of the Great Flood of 1993 were estimated between $12 billion and $16 billion.[22]

The flooding of the Rhine, Mississippi, and Missouri rivers provided costly lessons on treating the natural flow of rivers as a pathological condition.

The management and policy changes begun after the 1927 flood have had a number of profound effects. One was to shift the cost of flood control and relief from the local to the federal level. Another was to encourage people, farms, and businesses to settle in vulnerable areas with the knowledge that they would be bailed out of trouble at taxpayer expense. Since 1930, the U.S. Army Corps of Engineers alone has spent more than $25 billion on navigation efforts and flood control in the Mississippi basin. Billions more have been spent by other federal, state, and private interests.[23]

Numerous other subsidies from the federal government include providing crop insurance and crop price guarantees, paying for 80 percent of the cost of levees, and allowing farming within the former river channel, thus converting public land to private holdings. (In one state alone, the average annual cost of the subsidy was $71 per acre, not including the levees.) In fact, farming the land in the former river channel is profitable only with regular federal payments for flood damage.[24]

In 1968 Congress created the National Flood Insurance Program to cover flood-prone areas that private insurers deemed too risky, thus fostering rebuilding in many of those same areas. As with many subsidies that run counter to sound economic and environmental principles, this expensive program benefits relatively few. Nearly half of the billions of dollars paid out in flood claims went to the repeat flood victims who account for just 2 percent of the policyholders.[25]

On the 3,969-kilometer Missouri River, subsidized engineering and upkeep have maintained navigation and shipping for over 50 years. When the river engineering was planned in the 1940s, it was expected that 12 to 20 million tons would be shipped annually on the Missouri River, but this forecast was never realized. At its peak 20 years ago, shipping reached 3.3 million tons, and traffic has declined ever since. Despite its location in the nation's breadbasket, less than 0.01 percent of the region's corn and 2 percent of its wheat are transported by water. Many now recognize that the amount of shipping along the Missouri does not justify public expenditures for that purpose. In view of these facts, a broad-based coalition promotes restoring the waterway to its natural functions as a means both to reduce flood losses and to increase recreational use. Recreation already generates, at very little cost, at least four times the revenue provided by navigation.[26]

After the 1993 flood an Interagency Floodplain Management Task Force recommended ending the nation's over-reliance on engineering and structural means for flood control in favor of floodplain restoration and management. It emphasized managing the river as a whole ecosystem rather than as short segments. These findings echoed the conclusions of an extensive study of aquatic ecosystem restoration published by the independent National Research Council in 1992.[27]

Throughout the huge Mississippi River basin, the loss of forests, wetlands, oxbow lakes, and backwaters to settlement, agriculture, and river engineering has been extensive. The Missouri, for example, has been made shorter, deeper, and much narrower, and 95 percent of its floodplain has been converted to row crop agriculture; and the diversity of habitat types has shrunk as well. Separating fish from their floodplain spawning grounds and upstream reaches has virtually eliminated some species and caused many others to decline. The commercial fish catch has fallen 83 percent over the past 50 years. Harvests of the ancient paddlefish, for example, numbered 16,000 at one location around 1960; 20 years later only 125 were caught.[28]

In addition to removing the "flood-pulse advantage" for the river's fish, flood control and navigation structures have also adversely affected the integrity and productivity of the Mississippi Delta and the Gulf of Mexico. Because these structures trap sediments rather than allowing them to be carried downstream to replenish the delta as they have done for millennia, the coastal areas are actually subsiding as water inundates wetlands and threatens coastal communities and productive fisheries. Louisiana, for example, has the second highest volume of commercial fish landings in the country (over 1.2 billion pounds valued at $264 million in 1989)—a catch of primarily wetland-dependant species such as brown shrimp, white shrimp, blue crab, and sea trout. Yet between the direct loss of wetlands to other land uses and the loss due to subsidence, this bounty is imperiled.[29]

By destroying not only the fauna and flora, but the very ecological integrity of the Mississippi River system, we are eliminating valuable current ecological services. We are also foreclosing future evolutionary pathways and options for adaptation to climate change. The Mississippi is an old river where species have had a long time to diversify and create complex assemblages. Its richness is reflected in the fact that it is home to almost all freshwater mussels of the U.S. and one-third of its fish species. It also shelters some of the most ancient lineages of freshwater fish such as gars, sturgeons, and paddlefish. (In fact, the entire paddlefish family has only two living species, one in the Mississippi and the other in the troubled Yangtze River in China.) During the ice ages the north-south orientation of the Mississippi allowed species to shift to warmer waters. Today it also allows the yearly migration of waterbirds, shorebirds, raptors, and songbirds between northern breeding grounds and winter habitats in South America. This north-south orientation will also play an important role in future evolution and adaptation to climate change.[30]

Restoring the floodplains and managing the river basin as an ecosystem as recommended by the taskforce and others would not only help reduce flooding, it would also help restore other ecological services such as cleansing water and recharging aquifers. In addition, restoring the river would help renew its aquatic life and the health of the coastal ecosystem.

Unfortunately, the lessons learned from mistakes made on the Mississippi and Rhine rivers are seldom applied when untamed rivers are being developed. In South America, for instance, the countries of the Mercosur—the Southern Cone Common Market (Brazil, Argentina, Uruguay, and Paraguay)—and Bolivia are currently considering re-engineering that continent's natural infrastructure. The aim of the proposed Hidrovia project is to create a 3,400-kilometer, year-round shipping canal to facilitate the development of the interiors of these countries. But the 20-year project threatens irreversible environmental changes on a grand scale. And critics from the region and around the world were further alarmed when construction began before full environmental impacts had been assessed and citizens consulted.[31]

As originally conceived, the Hidrovia plans called for dredging and straightening the Paraguay and Paraná rivers, and draining a huge portion of what is considered to be the world's largest and most pristine wetland. The Gran Pantanal wetland now covers about 200,000 square kilometers at the junction of the Amazon rainforest, the savannas, and the Chaco wetlands. The Hidrovia project would demolish the natural rock formations at the southern end of the Pantanal that control the outflow of water (a vital part of a wetland that receives less than 69 centimeters of rain per year). In the first year of Hidrovia's operation alone, the Gran Pantanal would lose 17 billion cubic meters of water—enough to supply the entire population of Brazil, or nearly 155 million people.[32]

Critics believe that the impacts on this unspoiled system will include extinctions among its more than 600 fish, 650 bird, and 80 mammal species; increased flooding; and the disruption of many local Indian and non-Indian communities. Changes in the basic hydrology—the distribution and flow of water—of the region will diminish its ability to maintain water quality and act as a sponge to prevent flood and drought. Much as the opening of the earlier highway into the Amazon did, the construction of this navigation route would fuel the expansion of agricultural lands in the region, further accelerating environmental problems. Opponents of Hidrovia also challenged the assumptions and conclusions in the economic evaluation used by governments as the project's rationale—criticisms that were endorsed by the Inter-American Development Bank, a potential source of funding for the scheme.[33]

The intergovernmental committee responsible for Hidrovia has responded to the protests with modified plans and assurances that no additional work will proceed without an environmental impact assessment and that procedures for ensuring public participation and review will be estab-

Separating fish from their floodplain spawning grounds and upstream reaches has virtually eliminated some species.

lished. If implemented, these changes would be steps in the right direction. However, there is an enormous amount of momentum behind the project from senior officials and businesses within the five countries to build this aquatic highway as the backbone of the new Southern Cone free trade agreement. If the project proceeds, it is likely to trigger fundamental ecological, hydrological, and social disruptions—a cascade of troubles that would be felt throughout the food web and the economy.[34]

All over the world, rich freshwater systems are threatened by "pharaonic projects," as they are called in Brazil—massive public works projects that, like the pyramids of Egypt, require sacrifices from an entire society but bring little or no return. Now a major feature on the natural and political landscape, dams are at once symbols and agents of change. But the problems they cause can run far deeper than the still waters held in their reservoirs—from loss of fish to deterioration of human health. The construction of large dams has skyrocketed in the past 50 years. (See Table 1.) Today, they number more than thirty-eight thousand; over half are in China. In addition to the large dams that constrict the world's rivers, there are many smaller ones. The U.S., for example, has nearly fifty-five hundred large dams, and over one hundred thousand small ones.[35]

The rationales for building dams include hydropower production, flood control, irrigation, and storing water in the wetter season for use in the dry season (for irrigation, navigation, direct human consumption, etc.). Many were built before any environmental or socio-economic assessments were made. With the benefit of hindsight, numerous problems have been recognized.[36]

Dams provoke many of the same problems created by other water engineering and navigation works, reservoirs, and irrigation schemes. Usually, one of the earliest problems created by dams is the displacement of people from the area destined to be inundated by the reservoir. Nearly 5 million people have been or will be uprooted by dam reservoirs alone. Over half of those are in China. The forced removal of roughly 1.5 million people from the 600-kilometer-long reservoir of China's Three Gorges Dam will exceed any other dam-related human dislocation by an order of magnitude.[37]

Even when people are not directly displaced by dams, other severe human consequences can ensue. One is the "paradox of malnutrition" that frequently follows when irrigated (often monocrop) agriculture replaces traditional agrarian systems. In many regions people have developed flood-dependent agrarian systems over generations that include a mix of flood-recession agriculture, fisheries, herding, and natural products collection, producing a broad and reliable array of foods. In such places, small or large changes in the natural functioning of rivers can literally make the difference between life and death. Ironically, diversified flood-dependent systems produce more per unit area than irrigated agriculture. The dynamics of aquatic systems and sustainable human use are often poorly understood by outsiders. Thus, when freshwater systems are disrupted by engineering, ensuring that critical resources are maintained for the people who need them is a difficult proposition.[38]

Dams, reservoirs, and irrigation schemes can have other major impacts on human health unforeseen by project planners. Numerous studies have found that many large water engineering projects have directly increased the incidence of disease. For example, water resources development in the Senegal River Basin has resulted in epidemics of bilharzia (also known as schistosomiasis) and rift valley fever in areas that had previously been unaffected by these debilitating diseases. And malaria cases have proliferated as mosquito vectors found many new breeding sites. The incidence of cholera and other diarrheal diseases has grown as well.[39]

Many of the problems created by dams and other engineering structures result from the fact that they induce numerous fundamental ecological and hydrological changes. They change the dynamics of river ecosystems, fragment existing systems, and join previously unconnected ones. Dams alter the temperature and flow regimes of rivers; they are also barriers to migrating organisms such as fish, and to the natural movement of sediments, nutrients, and water—all of which feed the surrounding floodplains and ultimately the sea.

By reducing the flow of fresh water, dams and diversions can lead to the intrusion of salt water into previously fresh surface water and groundwater—rendering them undrinkable. Reducing the movement of natural sediment loads out to the sea causes the recession of fertile coastal deltas. The changes wrought by dams can precipitate declines in both freshwater and marine fisheries. In one of many well-documented cases, dams in northern Nigeria have lowered catches by over 50 percent and changed the species composition of those fish remaining. Egypt's Aswan High Dam, which impounds 50 to 80 percent of the Nile River's flow, also contributed to sharp drops in fish populations in the eastern Mediterranean Sea.[40]

As with other types of anthropogenic disturbances, many of the most serious problems that result from dams do not emerge for years or decades, and many are unanticipated. For instance, the reservoirs created behind dams have distinctive aquatic environments that are inhospitable to most native species. While dam proponents often claim that reservoirs increase fish production (which sometimes holds true for short periods following construction due to the proliferation of opportunistic or artificially stocked species), a marked decline takes place in all species as the reservoir environment system stabilizes. Furthermore, the same amount of fish could be produced in a natural river system in one-half to one-third the time, thus allowing other productive activities during the remainder of the year.[41]

Dam building can cause additional problems, both biological and institutional. In the tropics, higher siltation rates than that in cooler zones means the dams have a relatively short operating life. When dams silt up before they are paid for—with power decreasing over the years—they are even more economically risky than those built in a temperate climate. Globally, 75 percent of hydropower dams exceed their estimated cost, and many operate at less than their expected capacity, thus generating less than the promised return. The large amounts of rotting submerged forests in the reservoirs can emit more greenhouse gases (such as methane and CO_2) than coal-fired powerplants, thus weakening the rationale that tropical dams will help ease global warming. When the large start-up, operating, and maintenance costs of irrigation and dam projects are added in, they frequently make little ecological, social, or economic sense.[42]

In many places, chronic financial irregularities, corruption, and lax building standards and enforcement can cause serious financial and safety problems. Dam-related catastrophies are not without precedent. Secret Chinese government reports recently uncovered by Human Rights Watch/Asia reveal that in 1975 dam collapses killed approximately two hundred thousand people and brought famine and disease to eleven million others. If the dam at Three Gorges were to break it would release a flood 40 times larger than the earlier disaster.[43]

Although the flow and floods of the Mekong may seem wasted to an engineer, the river, the fish, and the people already make up a highly productive system.

Many of the ecological problems associated with dams have become apparent in the Columbia-Snake River Basin in North America, the region with the longest history of hydropower development. Even after decades of retrofitting older dams and designing new ones, effective ways of allowing dams and fish to coexist have yet to be found. Today the number of wild salmon returning to the Columbia River is less than 6 percent of what it was before the dams were built in the 1930s, despite the assurances of engineers at the time that "no possibilities, either biological or engineering, ...[were] overlooked in devising a means to ensure perpetuation of the Columbia River salmon." In the face of all the cost overruns, as well as the social and environmental costs so long ignored, the successors of those engineers have begun to realize how false such promises were. According to Dan Beard, former Commissioner of the U.S. Bureau of Reclamation, the agency responsible for much of the country's water development, "it is a serious mistake for any region in the world to use what we did on the Colorado and Columbia rivers as examples to be duplicated."[44]

Unfortunately, however, that is just what is happening in the Mekong River Basin. In the 1940s and 1950s, engineers heady from building a series of record-breaking dams in the Columbia basin drew up plans for controlling Southeast Asia's Mekong River, which begins in China and drains Myanmar (formerly Burma), Laos, Cambodia, and portions of Thailand and Vietnam. During the protracted conflict in Southeast Asia, the plans lay idle. But now that transforming "the Indochinese battlefield into a market-place" is the stated priority of Thailand, there is no shortage of willing investors eager to profit from this populous emerging market rich in natural resources. Private foreign investors, the Asian Development Bank, and various bilateral and multilateral aid agencies are helping Thai officials and the dam builders promote a harnessed Mekong as the path to prosperity.[45]

Experience with other large river systems suggests that the construction of a staircase of dams and diversions along the Mekong will take a high toll on the region's ecological, cultural, and economic lifeline. They will alter the basic flow and hydrology of the 4,200-kilometer Mekong, affecting fisheries, agriculture, and water supply along its length. Diversion of the natural flow of silt-laden waters will also cause the rich Mekong Delta to recede, as it depends on the deposition of these sediments. When dam enthusiasts hold up the promise of taming the Mekong's floods, they fail to mention that flooding is a natural part of the river's complex cycle, essential to maintaining the health and productivity of the land, the river, and the sea.[46]

Ninety percent of the Mekong's flow comes during the monsoon rains between May and October, when the river swells to 40 times its dry-season volume. Fish migrate long distances between the river and the surrounding country-side during the flooded period, and then back to the river and estuary as floodwaters recede. It is estimated that 90 percent of the fish in the Mekong basin spawn not in the river but in the submerged forests and fields, where they also forage. The flood waters enrich agricultural fields as well. All in all, 52 million people depend on the Mekong for their food and livelihoods. In Laos and Cambodia, most of the population sustain themselves directly from the river. Although the flow and floods of the Mekong may seem wasted and out of control to an engineer, the river, the fish, and the people already make up a highly productive system.[47]

A vital part of that system is Cambodia's Tonle Sap, or "Great Lake." During the dry season, the Tonle Sap drains into the Mekong. But during the monsoons, the flow is reversed and the lake increases its size fivefold to 13,000 square kilometers—becoming the largest freshwater lake in Southeast Asia. This flooding enriches fields and supports the lake's four hundred fish species, most of which forage and spawn in the surrounding flooded forests. One of the most productive fishing grounds in the world, the Tonle Sap yields an annual fish catch of 50,000-60,000 tons—the major protein source for Cambodia's 9.5 million people. The lake is also home to thousands of birds, including the endangered Sarus Crane. Cambodia has nominated part of the Tonle Sap as a UNESCO World Heritage Site.[48]

Diverting water from the Mekong river basin would produce a cascade of impacts on the productivity of the Tonle Sap. Because of its flat and shallow topography, the great lake is extremely sensitive to any reduction in the flow or volume of water. Reducing the Mekong flood discharge by 50 percent (as planned with upstream diversions) would decrease the lake's flooded forest area by 3,000 square kilometers, and hurt the productivity of the region that produces half of Cambodia's rice and almost all of its fish. A mere 1-meter decline in the lake's water level during the wet season would diminish the water stored by 15 billion cubic meters, impairing its capacity to hold and slowly release water for the region.[49]

The fate of such places as Tonle Sap (and the people who depend on them) may now rest in the hands of a recently resuscitated international body called the Mekong River Commission. This group could have provided a forum for coordinating truly sustainable river use and investment by its signatory countries—Thailand, Laos, Cambodia, and Vietnam. But in view of the weakness of its new rules and the momentum behind the dam projects there is little hope that either the cooperation or the sustainable development of the Mekong River promised in an agreement signed in 1995 will be achieved.[50]

Under the Commission's old rules, member countries could veto diversion or dam projects on the Mekong or its main tributaries. But the new rules ask only that the Commission be "informed" of such plans; and they fail to provide redress of damages—setting the stage for serious problems in the less powerful countries. Plans and contracts for building the dams are already underway, despite the fact that detailed assessments have not been made of the ecosystem and its needs, and despite the fact that this course will ignore the rights and needs of small-scale subsistence resource users, who form the majority of the population.[51]

The Commission's new rules are enthusiastically supported by the government of Thailand, the region's economic giant. Its demand for electricity is increasing by 10 to 15 percent a year and it has overtapped its own water resources, but it has less access to the Mekong than most of its neighbors. Imports from huge hydro developments it is promoting in Laos and Cambodia could give Thailand cheap electricity and abundant water—and allow it to avoid much of the attendant cost and damage. Similar projects the government is promoting in China, which is not a signatory country, would allow it to bypass the Commission (and public scrutiny) altogether.[52]

Power and water imports might also let the government sidestep its growing chorus of critics at home. Relative freedom of speech in Thailand has allowed people to air their concerns about the destruction of forests and fishing grounds, and the social impacts of relocating villages. The country's neighbors, on the other hand, have little press freedom. There has been scant public debate in Laos of the government's decision to feed Thailand's energy appetite by building more than 23 dams by the year 2010. Few of the people who will be forced out of their self-sufficient communities— whether by submerging or drying up their land, or by destroying their fish resources—have a hope of influencing these plans. But a broad spectrum of citizen, environmental, and human rights groups have opposed the new agreement. Many scientists, development planners, and hydrologists have also questioned the plans on technical grounds.[53]

> **Though the dams will cease functioning in a few decades, people will continue to pay for them for centuries.**

The conclusion of an editorial in a Thai newspaper that "Laos may be rushing after fool's gold" seems even truer now that Thailand's most recent energy forecast may mean a 50 percent reduction in the amount of power that they promised to purchase (they are also trying to negotiate a reduction in the price for the energy they do buy). Though the dams will cease functioning in a few decades, the people will continue to pay for them for centuries.[54]

The Pak Mun Dam in northeast Thailand provides a glimpse of a few of the problems likely to arise in the future. Sited on the Mekong's largest tributary in Thailand, the dam was built during the early 1990s on the strength of cursory environmental impact assessments that did not study seasonal fish migrations or the relationship between the river and the local people. Villagers were not asked about their use of the river, nor were they given access to information on the project—they were not even given maps, which would have let them see what was planned. Despite lengthy protests, more than two thousand families were evicted and thousands of others lost their source of food and income. Predictably, since the project's completion in 1994, all 150 fish species have virtually disappeared from the Mun River. In 1995, Thai officials admitted this destruction, but villagers are still struggling to receive compensation.[55]

The planned dams in the Mekong region have already had an unexpected effect—accelerated deforestation. As it is, logging—both legal and illegal—is difficult to regulate in a region with few roads, weak forestry departments, and strong vested interests. The illegal trade in drugs, timber, and gems is well known. With the possibility that forests may be submerged by the dams, the race is on to log large areas. Eyewitnesses report that trucks loaded with valuable virgin pine from Laos stand bumper to bumper at the Laos-Vietnam border. The effects of soil erosion and sedimentation from deforestation and mining activities have already been felt in the region. In Cambodia, logging has so far reduced the forest cover from 73 percent to 40 percent in the past 25 years. Two timber concessions were recently signed by Cambodia's prime minister in violation of his nation's log export ban and forest management rules. These two concessions alone allow 30 percent of the remaining forest to be logged by Malaysian, Indonesian, and Thai timber companies.[56]

If today's rapid resource extraction continues and the dam building proceeds as planned, the largest and most productive natural river system in Asia will be irrevocably altered. While it may bring what a banker in the region terms a "bonanza for construction companies, bankers, and engineers," it will push people who are subsisting on the edge into deep poverty. The very people with the most expertise in sustainably managing the river and the greatest need for its bounty will be cut off from it.[57]

What We Take Out, What We Put In

Increasingly, freshwater systems must compete with human beings for the very basis of their existence—water. Indeed, much of the water engineering described thus far is in support of increasing human appropriation of water resources. Water is diverted from rivers, lakes, springs, and aquifers for irrigation, flood control, domestic consumption, and urban and industrial uses. The amount of fresh water withdrawn has risen 35-fold in the past three hundred years; over half of that increase has occurred since 1950. (See Table 1.) Today, we withdraw water far faster than it can be recharged—unsustainably mining what was once a renewable resource.[58]

There are numerous examples of water overconsumption harming aquatic systems and thus human well-being. Diversions from the Nile River and sediments trapped behind its dams have caused the fertile Nile Delta to shrink and 30 out of 47 commercial fish species to become economically or biologically extinct. So much water is taken from the Colorado and the Ganges rivers that little if any reaches the sea, with harmful effects on coastal habitats, fisheries, and peoples. Lake Chad, in Africa's Sahel region, has shrunk by 75 percent in the last 30 years due to drought and diversion for agriculture, to the detriment of this important inland fishery and the people reliant on its diverse resources. In the former Soviet Union, diversion projects on some of the major rivers have reduced fish catches by 90 percent.[59]

Central Asia's Aral Sea is one of the most graphic examples of excess diversion and gross mismanagement of water resources. Since 1960, this massive body of water, once the world's fourth largest lake, has shrunk by half and lost three-fourths of its volume. Today, 94 percent of the river flow that once fed the Aral Sea is diverted to irrigate thirsty crops such as cotton in this arid region. Meanwhile, the sea's salinity levels have tripled and 20 of its 24 fish species have disappeared. The fish catch, which once measured 44,000

tons and supported sixty thousand jobs, is now non-exis-
tent. Fishing docks sit in the dust tens of kilometers from
the shrunken shoreline. Over 36,000 square kilometers of
former lake bottom are dry and bare. Severe dust storms
carry toxic salts and dust thousands of kilometers and afflict
three-quarters of the region's 3.5 million people with serious
illnesses.[60]

Aquatic habitat has disappeared due to excessive water
withdrawal for human use in other places too. In North
America, the southeastern U.S. and arid areas of Mexico and
the southwestern U.S. have especially high concentrations
of aquatic species in jeopardy. California has lost 91 percent of its over 2 million hectares of wetlands; 63 percent of
its native fish are extinct, endangered, threatened, or declining. Sixty-four percent of the Colorado River's water is withdrawn for agriculture (much of it for use outside of the river
basin) and 32 percent evaporates from the reservoirs. With
such heavy withdrawal and loss, little wonder that the river
bed is often dry before it reaches the sea and that its fish
species are among the most endangered in the world.[61] (See
Table 2.)

Humans have always relied on water systems to carry
away their wastes, but the increased load of what we put in,
exacerbated by the loss of water we withdraw, has reduced
the capacity of rivers to assimilate or flush pollutants from
the system. Agriculture is not only the biggest consumer of
water (see Figure 3), but also the biggest polluter. What little water is finally returned to streams and rivers after irrigation is severely degraded by increased toxicity, excess nutrients, salinity, higher temperature, pathogen populations,
sediments, and lower dissolved oxygen. Excess nutrients,
especially nitrogen and phosphorus coming from human
and animal wastes and fertilizers overstimulate the growth
of algae. As the overabundant organic material dies and
decomposes, it robs the water of life-giving oxygen, a
process called eutrophication. One of the most widespread
forms of pollution, this loss of dissolved oxygen can be
deadly for fish and other aquatic organisms.[62]

FIGURE 3

Global Water Withdrawal by Sector, 1900–2000

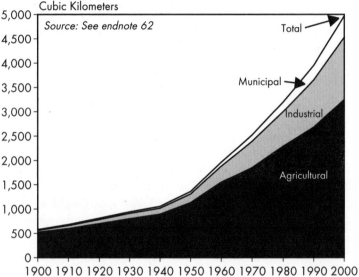

Chemical pollutants are also released directly from industrial and municipal sites, and indirectly as runoff from homes, roads, and cities. Toxic chemicals are now ubiquitous in the global environment. Their volume has increased substantially and they are very persistent. Not all pollution is dumped directly from the end of a pipe or even produced within an affected watershed; much is carried long distances by air currents.

One widely studied form of air pollution is "acid rain." Because most aquatic organisms are adapted to a narrow pH range, changes in the acidity of water can be fatal. For example, when acid precipitation causes the water's pH to fall below 5.5, salmon embryos die and hatchlings develop abnormally. Adult fish are affected too. They are stunted in size, which diminishes their chances of making the arduous upstream migration, and decreases the number of eggs laid if they do survive the trip. In addition to the direct toxic effects, low pH causes toxic metals (such as aluminum, mercury, and lead) to be leached from soils, with fatal consequences for fish, plants, insects, and invertebrates. Changes in pH also cause changes in nutrient availability, with obvious implications for the food web. Acidification is especially severe in the northern hemisphere, in industrialized areas, and regions downwind.[63]

The world's largest freshwater ecosystem, the Great Lakes of North America, has already felt the full range of anthropogenic stress. The region is now home to more than 38 million people—and to significant portions of North America's industrial and agricultural activity. Over the last two hundred years, the Great Lakes basin has lost two-thirds of its once extensive wetlands. Barriers, canals, dams, and channels have eliminated vast fish spawning grounds. Today, less than 3 percent of the lakes' 8,661 kilometers of U.S. shoreline is suitable for swimming, for supplying drinking water, or for supporting any aquatic life.[64]

Pollution is a major cause of the plight of the Great Lakes, which hold 20 percent of the world's fresh water. The addition of agricultural runoff, human waste, and household detergents over more than a century have caused extensive eutrophication. One early attempt to deal with Great Lakes pollution was construction of the Chicago Sanitary and Ship Canal, rerouting waste from Lake Michigan into the Mississippi basin's Illinois River in 1900. While easing Chicago's pollution problem, the canal dealt quite a blow to the river by causing severe eutrophication. In more recent decades a shift in agriculture from soil-covering crops (such as forage and pasture) to row crops has greatly increased soil erosion and sediment load in the region's rivers. Since 1908 the commercial catch in the Illinois River has fallen 98 percent; the river that once produced 10 percent of the U.S. freshwater fish catch is now virtually barren.[65] (See Table 2.)

Today the most troubling source of pollution in the Great Lakes is toxic chemicals. Despite the improvements brought about by decades of regulation in the United States and Canada, huge quantities of these pollutants are still introduced into the lakes every year. And the toxics from previous years tend to remain in the water and bottom sediments because the basin is a relatively closed system—only 1 percent of its water flows out annually. Fifty to 100 million tons of hazardous waste are generated around the lakes. Over 25 million tons of pesticides alone are used annually in the Great Lakes watershed. The large surface area of the Great Lakes makes them vulnerable to deposition of airborne pollutants, which today account for the majority of the most toxic substances entering the system. Some of the contaminants blow all the way from farms in Mexico and cement plants in Texas, provoking fish consumption advisories even in the relatively pristine Lake Superior, the world's largest lake. In the form of acid rain, airborne deposition is killing many of the eighty thousand smaller lakes in the region and putting growing stress on the Great Lakes themselves. The Great Lakes' "airshed" is even larger than its 520,590-square-kilometer watershed.[66]

The health and composition of lake fish offer a good clue to the health of the whole system.

The health and composition of lake fish offer a good clue to the health of the whole system. In 1993, two-thirds of the nation's 1,279 fish consumption advisories were issued in the Great Lakes region, mostly due to the presence of mercury, PCBs, chlordane, dioxins, and DDT. With more comprehensive data, the picture would likely be much worse; as it is, only 1 percent of the thirty thousand different chemicals entering the lakes are reliably monitored.[67]

The lakes' toxic brew sometimes causes massive fish kills, but the more subtle effects may be even more dangerous. Many of these chemicals become more concentrated as they pass up through the food web, with top predators—such as humans—receiving the highest doses. This process is known as bioaccumulation and biomagnification. A person would need to drink Great Lakes water for more than a thousand years to take in the amount of PCBs one would ingest eating a two-pound trout.[68]

Many of the compounds—such as DDT, PCBs, agricultural chemicals, even some components of detergents and plastics—act as endocrine disrupters by mimicking the action of hormones. In very minute amounts, they alter a whole spectrum of morphological, physiological, reproductive, and life history traits. Tumors, deformities, reproductive abnormalities, and reduced survivorship are widespread in exposed fish, birds, and mammals. A 50 percent decline in human sperm counts worldwide since 1940 (when widespread use of these chemicals began) has been attributed to the ability of many of these substances to mimic estrogen, the feminizing hormone. And endocrine disrupters assimilated by one generation can produce changes in the next. The cognitive, motor, and behavioral development problems noted in children of women who eat contaminated fish do not come from what was eaten during pregnancy, but from what was consumed by the mother throughout her life.[69]

Attempts to heal the Great Lakes have been limited largely to episodes of crisis management, often with dubious results. When human epidemics were traced to the release of sewage, outfall pipes were extended farther into the water. When the native fisheries collapsed, exotic species were stocked in their place. And most pollution control has focused on end-of-the-pipe management of individual chemicals rather than on comprehensive source reductions.

Still, some progress has been made. The International Joint Commission, the Great Lakes Fishery Commission, and the Great Lakes Water Quality Agreement (GLWQA) of 1978 have provided forums for binational cooperation—on both federal and state levels.   Cooperation under the GLWQA has significantly reduced phosphorus levels and eutrophication.   In the United States, federal and state authorities have substantially reduced inputs of many pollutants and have developed flexible guidelines for achieving further reductions. A new five-year strategy developed by federal, state, and nongovernmental agencies and coordinated by the U.S. Environmental Protection Agency (EPA) takes an integrated, ecosystem management approach to problem solving and decision making in the region. Although it is too soon to judge its success, the strategy represents an important conceptual leap from the days when a crisis-by-crisis response held sway.[70]

Fisheries Mismanagement

Some of the most well-documented evidence of declining freshwater systems comes from commercial fisheries data. But a quick look at the global trends in freshwater fisheries—where the catch has either remained stable or increased slightly over the past 25 years—belies the complexity of the situation. The United Nations Food and Agriculture Organization (FAO), which compiles much of these data, considers most inland fisheries to be exploited at or above sustainable levels. Indeed, many once abundant fisheries have been eliminated.[71] (See Table 2.)

Overexploitation of freshwater fish by commercial, subsistence, or even recreational fishers can seriously alter the abundance and diversity of fish species. While fishing is a rather visible activity, it is not the primary cause of the *global* decline in freshwater species or systems. Combined with other stresses however, overexploitation and mismanagement of fisheries can lead to the collapse of a region's fish fauna. Sometimes overfishing constitutes the initial stress, sometimes the final fragmenting force.

River fisheries respond to stress in a predictable sequence, regardless of the source, be it overfishing or habitat modification. According to fisheries biologists, the first sign of stress is the loss of large and migratory species. Once this happens, the overall catch weight can remain level for a while, albeit comprised of smaller fish. Exotic species may invade (or be intentionally added to) the system and displace the remaining native species. As pressures from with-

TABLE 2

Status of Selected River Systems and Lakes—
Evidence from Commercial Fisheries

China: 80% of China's 50,000 kilometers of major rivers are so degraded they no longer support fish. *Causes:* pollution, watershed degradation, water withdrawal, overfishing. **Yangtze River:** 70% of China's catch once came from this river. There were 50 species of economic importance in the 1960s, 20 in the 1980s. Annual catch fell from 458,000t in 1954 to 226,000t in the late 1970s and has continued declining. (The 3 Gorges Dam will cause further degradation.) **Huang-he (Yellow) River:** Large species were 72% of catch in 1950s, only 21% in 1982. Annual catches of all species in the lower reaches fell from 5,000t in 1950s to under 150t in early 1980s. **Amur River:** In 1969 large species were 69% of the catch, in 1982 only 10% and they averaged 30% smaller.

Mekong River (Southeast Asia): In Vietnam, conversion of 1,000 km2 of mangrove swamps in the Mekong Delta has caused a 2/3 drop in its inland fisheries alone. Cambodia's Tonle Sap (Great Lake) has had a 1/3 drop in fish production from 100,000t per year in the 1960s. *Causes*: deforestation, intensification of fishing and agriculture in the watershed. (A series of planned dams along the Mekong and its tributaries and water diversions from the basin will severely damage the health and productivity of the river and the lake.)

Rhine River (Europe): Of 44 fish species, 8 have been eliminated, 25 are rare or endangered. *Causes*: river confined to a single channel, cutting off the river from floodplain; loss of gravel beds for spawning; heavy pollution.

Oueme River (Benin, West Africa): Severe decline in quality and quantity of catch from the Oueme floodplain. Loss of 4 of 93 freshwater fish species by 1967, many others less abundant, smaller in size. *Causes:* overfishing, population pressures, agricultural development, compounded by the 1980s drought in the Sahal.

River Senegal (West Africa): Fish production has dropped by over 50%. *Causes:* damming the river and the Sahaleian drought.

Lake Victoria (East Africa): 200 endemic fish species are extinct, the remaining 150 are endangered. *Causes*: predator species (such as the

Nile perch) introduced since the 1950s. In 1976 only 0.5% of the commercial catch were exotics; 7 years later exotics were 68% of the catch.

Magdalena River (Colombia, South America): Drop in fisheries production from 72,162t in 1977 to 23,321t in 1992 (over 2/3 decline in 15 years). *Causes:* heavy fishing of large migratory species, pollution from oil production, land use changes.

Colorado River (USA): Has 49 native fish species, 86% endemic. 40 of the 49 are now either at risk of extinction or already extinct. *Causes:* extensive system of dams and reservoirs and excessive withdrawal of water from the basin. Additional pressure from 72 introduced species including many predatory fish.

Missouri River (USA): 83% drop in commercial catch since 1947 (loss of 216,000t of fish per year); what remains today is in short unchannelized portions of the river. *Causes:* river channelized and impounded for navigation; row crop agriculture in 95% of the floodplain.

Illinois River (USA): In 1908, commercial yield of fish was over 10,000t per year (equal to 10% of the U.S. freshwater catch), and employing 2,000 commercial fishermen. By the 1950s fisheries had dropped 98%. Similar trends in sport fish and waterfowl populations. *Causes:* heavy sewage loads and eutrophication early in the century, heavy soil erosion from intensified agriculture in recent decades.

North American Great Lakes: In 1900, native salmonids (trout, etc.) comprised 82% of the commercial catch, by 1966 only 0.2%; the remaining 99.8% were exotic species. Current status of the 11 species of native fish that once provided a commercial catch of 1.4 million kilograms per year: 4 extinct, 7 at risk. *Causes:* extensive land use changes, pollution, exotic species.

Aral Sea (Central Asia/Former Soviet Union): 20 of 24 fish species disappeared, along with the commercial fishery that supported 60,000 jobs and a catch of 44,000t Lake has lost 3/4 of its volume, salinity levels have tripled. Toxic salts blowing from 14,000 square miles of dry lake bottom have afflicted 75% of the region's people with serious illness. *Causes:* 94% of the river flow into the lake is diverted for irrigation.

Source: See endnote 71

in and without continue, catches fluctuate and eventually decline. The system and the fishery fall into a downward spiral, and more and more human intervention is required to prevent their total collapse. Once they have crossed the "stress threshold" it is very difficult to revive them.[72]

In most industrialized countries the natural freshwater fisheries have already collapsed due to the cumulative pressures of overharvest and other assaults on the system. Most of the fish consumed there today comes from aquaculture and marine sources. While per capita consumption is high, fish are generally not a major source of protein (between 7 and 10 percent of total animal protein in North America and Europe, for example).[73]

In most of the world, however, fish are a critical food source, providing a significant proportion of animal protein for over 1 billion people. In Africa, fish provide 21 percent of total animal protein, in Asia, 28. In many countries it is much higher. For many of the world's poor, continued access to a self-sustaining natural fishery is literally a life or death issue. In land-locked countries with no access to coastal and marine fisheries, these resources are even more significant. International statistics do not reflect their full importance, because many of the fish that are caught never enter the formal economy. They are used by fisher families or traded locally. Fish are an important "free" and abundant resource, and exploitation by subsistence fishers is usually at very low levels, staying within the sustainable capacity of the local resource.[74]

At the same time the subsistence value of fisheries is rising, the international trade in fish is expanding rapidly. Herein lies part of the fisheries' dilemma. Fish are an increasingly valuable export product, particularly for developing countries, far surpassing other major commodities such as coffee and rice. Nearly half of the $40 billion fish trade comes from developing countries. Thailand, for example, is now the world's largest fish exporter (and the top shrimp-farming country). Orienting economies and resource management to export- and infrastructure-intensive industries may be financially attractive in the short run. But too often "putting all the eggs in one basket" subject to whims of international markets has provoked numerous fiscal, social, and ecological disasters.[75]

The combination of commercial overfishing and channelling fisheries' output into exports means that fish are becoming less available and less affordable for those that need them most. In Bangladesh, for example, per capita supplies have fallen off by nearly one-third in the past 20 years. Despite this drop, its export-oriented fish culture has expanded rapidly: Bangladesh is currently the 11th largest aquaculture producer in the world. Fish, often considered the food of the poor, are floating further from their grasp.[76]

Around the world, artisanal fishers—those fishing for subsistence and local markets—and commercial fishers increasingly compete for the same catch. In a pattern seen in both freshwater and coastal marine fisheries, overexploitation has often followed in the wake of a shift from small-scale artisanal fisheries geared to subsistence and local markets to intensive commercial exploitation for distant markets. Commercial fishers usually do not know or respect the local customary restrictions on fishing that have helped keep it sustainable. In an all too familiar pattern, the benefits of overexploitation accrue to one group, while the costs are borne by another.[77]

For many of the world's people, continued access to a self-sustaining natural fishery is a life or death issue; in land-locked countries, these resources are even more significant.

What is happening to the Amazon River shows how even an immense body of water can be overstressed, especially when unchecked commercial exploitation takes hold. Logging, mining, dam construction, and conversion of the floodplain for agriculture and livestock have been rapidly expanding. In the lower third of the Amazon, only 15-20 percent of the flooded forests remain. Flowing 6,500 kilometers to the Atlantic Ocean, the Amazon carries one-fifth of the fresh water that flows into the world's oceans. It is home to over three thousand species of fish alone—one-third of the world's total. During half of the year, portions of the forest up to 20 kilometers from the river are naturally flooded by several meters of water and become even more productive than they are during the drier season. Three times as much fish is harvested from flooded forest as from the river itself. The majority of the 200,000 tons caught each year in the basin are pulled in by small fishers for subsistence and local market consumption. Around major river towns, however, a shift from subsistence fishers to unsustainable exploitation by commercial fishers is contributing to the decline of many species. As a result, some of the most popular food fish, such as the 30-kilogram *tambaqui*, are already becoming difficult to find.[78]

Underlying the overexploitation of many species is the inability of humans to accept and respect the natural limits of aquatic ecosystems and to recognize the impacts of their own actions. People have expectations that harvest levels set during earlier periods of abundance can remain the same despite natural fluctuations and declines caused by overharvest and other human activities.

Once the problems caused by interfering with the natural functioning of ecosystems are recognized, human attempts to rectify them often do more harm than good. As river fish respond to stress in a predictable way, so do humans to declining fisheries. Human reactions include: switching fishing effort to another species, intensifying fishing effort (often using increasingly destructive means), artificially supplementing water bodies with hatchery-bred or exotic fish, and shifting to aquaculture. It is a rather fragmented species-by-species approach to management similar to those agriculture and forestry production methods that seek to maximize current commodity production with little regard for long-term sustainability or the health of the system. Most of the management effort has been expended on activities such as hatcheries and aquaculture that effectively postpone the perceived need to maintain natural systems more properly.

One common management response to declining natural fish populations is to raise young fish in hatcheries and release them in the wild. But contrary to expectations, adding hatchery fish does not actually increase the number of adult fish that can be caught. The release of hatchlings brings in new diseases, additional competition, and fish that are smaller and less fit than wild fish. It can also weaken the genetic base of wild fish through interbreeding. In the Pacific Northwest, for example, hatchery-raised salmon have brought many native stocks to the brink of extinction while hiding the decline of wild salmon and delaying real remedies.[79]

Often the preferences of politically or economically powerful users may determine how the fishery is managed, which species are maintained, which are stocked, and which are suppressed. In the tropics, the unbridled conversion of wetlands to aquaculture ponds for shrimp production in recent years has provided substantial short-term economic benefits for a powerful minority while the resultant loss of fisheries, livelihoods, fresh water, and flood control is paid for by a powerless majority.

A major recent response to the diminishing catch from inland fisheries is the development of industrial-scale aquaculture, where fish, shellfish, and plants are grown in high concentrations in confined areas. Since 1984, when the FAO began collecting separate aquaculture statistics, output has nearly doubled. Aquaculture now supplies 14 percent of the global fisheries production (marine and freshwater). The global industry, valued at $27.6 billion per year, is expected to continue expanding.[80]

The rapid expansion of aquaculture and the increasing importance of fisheries as a source of export revenue have had a number of consequences. China, for example, expanded its fisheries production 300 percent from 1982 to

1992. Today it provides over half of the world's total inland water catch and nearly half of all global aquaculture output. However, behind this rosy picture lie more telling statistics. Increases are due not to better fisheries management but to serious overexploitation. Eighty percent of the country's 50,000 kilometers of major rivers have been degraded to the extent that they can no longer support fish. Significant overexploitation of the rivers that still support fish and intensive aquaculture operations have been propping up the total output, but they are not likely to be sustainable. (See Table 2.) Despite these ominous signs, China has announced its intention to intensify and expand the acreage converted to aquaculture in order to increase its production 50 percent by the year 2000.[81]

The increase in the aquaculture "catch" also hides the decline of natural fisheries and the role played by aquaculture itself in degrading freshwater and coastal systems. In Vietnam, for example, the conversion of 1,000 square kilometers of mangrove swamps to shrimp culture ponds is believed responsible for a two-thirds decline of the Mekong Delta's rich capture fisheries. Similar ecological declines have occurred in other countries that have converted large areas to aquaculture over the past decade (such as Thailand, Ecuador, Bangladesh, and Indonesia).[82]

The impacts of aquaculture stem from the fact that these operations consume large quantities of resources and produce vast quantities of waste. They also degrade the integrity of the aquatic systems that support them. As vast areas are converted, important ecosystem services are lost, rendering systems unfit to serve as nurseries and feeding grounds for native aquatic species. In many cases, the aquaculture operations are abandoned after a short time once the environment deteriorates or they are no longer profitable, leaving the aquatic equivalent of a strip mine in their wake.[83]

Like other intensive livestock operations, organisms farmed at such intense concentrations are very susceptible to disease. In turn they use large amounts of chemicals and antibiotics. These chemicals can enter the wild food chain and alter aquatic microbes, by breeding resistent bacterial strains for example. There have already been numerous cases of hatchery and aquaculture operations introducing and spreading deadly new diseases into the aquatic environment. And as in other livestock operations, production is based on a limited number of species, with a relatively narrow gene pool, increasing its vulnerability to catastrophic loss. Nearly 80 percent of freshwater aquaculture production relies on carp species.[84]

The organisms raised can themselves escape into the wild environment. There are many examples of aquaculture species that have escaped confinement and destroyed native species as well as large expanses of habitat. Nearly a third of introduced fish species in European waters have come from aquaculture operations. The very attributes that make aquaculture species so well adapted to intensive production—high rates of growth and reproduction, ability to withstand a wide range of conditions, and so forth—make them invasive and destructive in the wild.[85]

Alien Invaders

The incursion of non-native, or exotic, species is another common if little known factor in the loss of freshwater species. According to the American Fisheries Society (AFS), it is cited in 68 percent of the North American cases. Exotics may prey on native fish, compete with them for food and breeding space, disrupt food webs, and even introduce new diseases. The spread of these species is a global phenomenon, one that is increasingly aided by the growth of aquaculture and by shipping and commerce.[86]

The growing problem of exotic invasions occurs across the landscape but it is especially serious in the aquatic environment. The list of these alien introductions, both intentional and accidental, is long and includes vertebrates (e.g., fish and mammals), invertebrates (e.g., mussels), higher plants (e.g., water hyacinth) and microscopic plants, and animals (e.g., spiny water flea, dinoflagalates). Most incursions are recent; for example, in the Northern Mediterranean, 60 percent of exotics have arrived in the last 40 years.[87]

Exotics can also mask ecosystem decline with significant implications for policymaking. Intentionally introduced exotic sport fish are a kind of fisheries sleight-of-hand. People seeing abundant fish may not realize these are not native and may not even be self-reproducing. In North America, for example, over four hundred species of freshwater fish have been introduced into a system outside their natural range. One hundred and forty species, including forty not even native to the continent, have become established. Two-thirds of the freshwater species introduced into the tropics have become established.[88]

Exotic plants can be especially harmful in freshwater systems, multiplying rapidly and absorbing large quantities of nutrients, shading out other species, obstructing the movement of people and boats, and providing a breeding ground for disease. One vivid example is the spread of the floating fern in the Sepik River floodplain of Papua New Guinea. In just eight years, a few plants expanded into a mat covering 250 square kilometers, seriously interfering with river transport and fishing for eighty thousand native people. The water hyacinth (discussed below) has been transported far and wide from its South American origins.[89]

One of the rarely recognized impacts of navigation is its role in the spread of exotic species. In earlier times, organisms might cling to the outside of ships, but because vessels moved more slowly then and the hitchhikers were subjected to a range of conditions, far fewer of the potential invaders survived. Today, however, large, modern ships not only move faster, they use vast quantities of water to stabilize themselves, and this ballast water is dumped at their destinations before they load cargo.

Transfers in ballast water occur in every direction along

all shipping routes, with serious economic and ecological consequences. About ten years ago, an Atlantic Coast comb jelly was accidentally transferred to the Black and Azov seas. Because it lacked natural predators in its new aquatic home, it gobbled up so much zooplankton that the once-productive Azov fisheries were shut down and the Black Sea anchovy fisheries were virtually eliminated, at a loss of $250 million per year. Little wonder that ballast water transport has been called a "giant biological conveyor belt" responsible for a "flood of irreversible global invasions."[90]

The Great Lakes of North America also suffer from the kind of "biological pollution" that the spread of exotic species represents. Two hundred years ago, each of the five Great Lakes had its own thriving aquatic community, with an abundance of members of the salmonid (salmon and trout) group. In 1900, native salmonids still comprised 82 percent of the commercial catch; by 1966, natives made up only 0.2 percent of the catch. The remaining 99.8 percent were exotic species.[91] (See Table 2.)

Most of the 130-plus exotics in the Great Lakes today have found their own way into the system by moving through canals or hitchhiking in ships. More than one-third of these aliens have entered the system in the 30 years since the opening of the St. Lawrence Seaway. The canals, for instance, are thought to have opened the Great Lakes up to the sea lamprey, an ocean-going parasite that devastated the lake fisheries. In Lakes Michigan and Huron, the lamprey is credited with driving the annual commercial lake trout catch from 5,000 tons in the early 1940s to fewer than 91 tons just 15 years later. The first effective intergovernmental cooperation in the region came about to combat the sea lamprey, and resulted in the Great Lakes Fishery Commission. The lamprey remains in the lakes and its control requires constant intervention.[92]

Prospects for controlling two more recent invaders, the minute zebra mussel and the quagga mussel, are far less certain. Inadvertently introduced to the lakes in 1988 from ship ballast water, the zebra mussel has already spread to most major rivers and lakes in the east, and has been found as far away as irrigation pipes in California. The larvae of this prolific Caspian Sea native attach to hard surfaces, such as rocks, other shellfish, pipes, and ship hulls. They form dense colonies on substrate used by spawning fish and native mussels and virtually eliminate the plankton needed by these fauna. Workers at Detroit's electric power generating plant have found as many as 750,000 mussels per square meter in the plant's water intake canal. Since an effective method for controlling them has yet to be devised, the cost to cities and industries of keeping these tenacious invaders from clogging intake pipes and heat exchanges could reach $5 billion by the year 2000 in the Great Lakes alone. By that time, the zebra mussel is expected to have colonized virtually all freshwater systems in North America.[93]

On the other side of the globe, one of Africa's Great Lakes is also suffering from the effects of an introduced species. The Nile perch is conspiring with changing land use, pollution, and growing population pressures to rob Lake Victoria of its rich fauna and its people of a valuable source of protein and employment. The Great Lakes in East Africa's Rift Valley are home to hundreds of small, colorful cichlid fish species that show an amazing array of colors, patterns, and behaviors. Because the three largest lakes are not connected to each other, and lie within different river basins, each one's fauna and ecology are distinct—99 percent of the fish found in each lake are endemic. Lake Tanganyika is the oldest and deepest, with the most diverse fish fauna of any lake on earth. Bounded by Uganda, Kenya, and Tanzania, the shallower Lake Victoria covers some 62,000 square kilometers. It is the largest of the Rift Valley lakes, and the second largest lake in the world.[94]

The introduction of the Nile perch to Lake Victoria in 1954—against the prevailing scientific advice of the day—has virtually eliminated the native fish population. This 200-kilogram predator more than 2 meters long consumes enormous quantities of little fish: since the perch was introduced, Lake Victoria has lost 200 taxa of endemic cichlids,

FIGURE 4

Impact of Introduced Species: The Demise of the Native Fishes of Lake Victoria

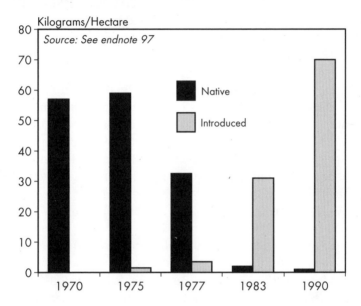

spectacular species found nowhere else; the remaining 150 or more are listed as endangered. The loss of Lake Victoria's fish is so severe that Boston University biologist Les Kaufman has described it as the "first mass extinction of vertebrates that scientists have had the opportunity to observe."[95]

On shore, a shift from little fish to big fish has taken place also. Until recently, the native fish of Lake Victoria were harvested by small-scale fishers and processed and traded by women for local consumption. This kept the fish's nutritional and economic benefits in the lakeside communities. Today, the perch are caught by large, open-water vessels with destructive gear, then processed and traded by big commercial operations for the export market. Local women are left with the scraps—which they must purchase. Deprived of work and unable to afford this higher-priced (and less palatable) catch, local people face a serious nutritional predicament. The perch takeover has decimated the primary economic and nutritional resource of 30 million people.[96]

The exact key to its success is uncertain, but the perch's known ability to change its lifestyle and breeding strategy to suit prevailing conditions may play a significant role. In the late 1970s, the lake's water began to undergo eutrophication. At the same time, the invader fish underwent a massive population explosion and quickly began consuming and displacing native fish. The results are apparent from fishery statistics. Kenya, for example, reported only 0.5 percent of its commercial catch as perch in 1976, but by 1983 the proportion reached 68 percent. While a small portion of that increase may be attributed to larger fishing vessels, more fisherman, and so forth, scientific fish surveys also show the demise of native fish and the takeover by introduced fish species.[97] (See Figure 4.)

It is unclear whether eutrophication gave the perch an opening, or whether this newcomer's consumption of native fish "decoupled" the lake's internal recycling and cleansing loop. Either way, the structure of the entire system has changed. Ten years ago, Lake Victoria was oxygenated to its bottom, 100 meters down. Now it supports life only in the upper 40 meters or less. Regular mixing events, in which the now suffocating, oxygen-depleted bottom waters rise to the surface, cause frequent fish kills. Ironically, the perch itself may now be in decline, from overfishing, periodic die-offs, and its own voracious appetite.[98]

The perch is not the lake's only alien species problem. Water hyacinth, native to South America, was first found there in 1989. Encountering no predators in Africa, the plant covers waterways quickly, depleting oxygen from the water and clogging intake pipes, irrigation canals, and ports. A single plant can blanket 100 square meters in just a few months, providing a breeding ground for disease vectors such as bilharzia-carrying snails and malaria mosquitos.[99]

There are other pressures on the ecosystem too. Millions of liters of untreated sewage and industrial waste flow into

Lake Victoria every day from Kisumu, Kenya's third largest city, and from Mwanza in Tanzania. Watershed degradation and agricultural runoff contribute chemicals, nutrients, and sediment. And from Rwanda came the grisly addition of forty thousand genocide victims that floated down the Kagera River in May 1994.[100]

Degraded and simplified, Lake Victoria is no more likely to make a stable "fish ranch" than are the North American Great Lakes. But the institutional challenges of caring for Africa's largest lake are nearly as complex as the ecological ones. A major cooperative effort among all three lakeside countries (Uganda, Kenya, and Tanzania), the Lake Victoria Environmental Management Program will focus on water quality, land use management, control of the exotics and water hyacinth, and community-based enforcement. Successful methods developed in pilot zones around the lake during the first few years will then be applied to larger areas. Such cooperation may yet restore Lake Victoria, and could also preserve the less-degraded Lakes Malawi and Tanganyika—the other jewels of the Rift Valley. The lessons learned about how to restore a system and how to use adaptive management—learning as you go and applying those lessons—could also help guide efforts in other parts of the world.[101]

Habitat Degradation

Each of the stresses discussed so far plays an important role in the loss of aquatic biodiversity and impaired ecosystem services. But taken together they are overwhelming, degrading and fragmenting the landscape. According to an American Fisheries Society study of North America, physical habitat alteration is implicated in 93 percent of the declines.[102]

Human changes to the landscape are extensive and accelerating. The stresses on ecosystems come not just from

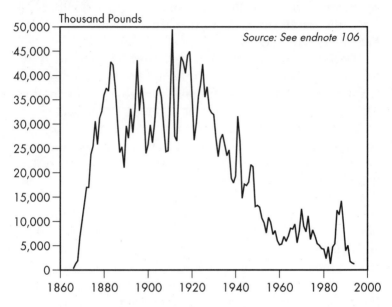

FIGURE 5

Commercial Catch of Salmon in the Columbia River, 1866–1994

the number of people but also their location and the nature and scale of their activities. Today, over 60 percent of the world's 5.6 billion people live within one kilometer of surface water, and virtually every human activity is ultimately reflected in the water.[103]

Dams, navigation channels, and flood control and irrigation structures are the most obvious signs of human intervention in the aquatic environment, but even in the absence of visible engineering works, the cumulative affects of human activities in the terrestrial landscape can be profound. Logging, mining, grazing, agriculture, industrialization, and urbanization all degrade rivers, lakes, and the lands they drain—the vital watershed—in ways that make them less able to support life and to provide valuable ecosystem services.

The loss of genetic, species, and ecosystem diversity resulting from habitat degradation and fragmentation is substantial. Entire regions may become devoid of many habitat types—such as wetlands, floodplains, waterfalls, and rapids—and the resident and migratory species adapted to those habitats. Many species require several different habitat types during different stages of their life histories. Eliminating habitats or preventing the movement of species, genes, materials, and nutrients among habitats has serious implications for species survival. Indeed, extinctions and extirpations of riverine species from large areas have already occurred. Many other species, whose populations have become too fragmented to remain viable, are likely doomed to extinction in the near future.[104]

The Columbia River Basin on the Pacific coast of North America clearly demonstrates the ways in which human stresses are cumulative and synergistic. Because of their unusual life history, salmon are good indicators of the health of this region's rivers and forests. Salmon hatch in streams and rivers, then make their way to the ocean. After several years, they return to their ancestral streams to mate, and often die. Not only do salmon form a symbolic link between the land and the sea, they also represent an important nutrient link as their carcasses feed terrestrial and aquatic organisms far inland.[105]

Around 10 million salmon a year once returned to this river basin to spawn in their ancestral streams; in 1992, only 1.1 million made it back, and most of them had been born in a hatchery. In 1895, nearly 20,000 tons of salmon and steelhead trout were harvested from the region's most important river, the Columbia. One hundred years later, the harvest was just 553 tons. In Idaho, one of the region's five Pacific salmon species—the coho—became extinct in 1986. In 1994, only one sockeye salmon, nicknamed "Lonesome Larry" by Idaho Governor Cecil Andrus, completed the journey from the Pacific Ocean up the Columbia and Snake rivers to Idaho's Redfish Lake.[106] (See Figure 5.)

Under the best conditions, just one of every five to ten thousand salmon eggs will be fertilized, hatch, and survive long enough to reproduce. And because these fish pass

through a wide range of habitats and conditions, over distances as long as thousands of kilometers, they are vulnerable to the full range of forces that nature and humankind
inflict. The Columbia and its main tributary, the Snake
River, drain 673,400 square kilometers in seven western
states in the United States and portions of the Canadian
province of British Columbia. Virtually no part has been
untouched by the forces implicated in the salmon's
decline—hydropower and other sources of habitat destruction, hatcheries, human population pressures, and overharvest.[107]

Today, the Pacific salmon's biggest physical obstacles are
the 58 hydropower dams and 78 multipurpose dams, and
the nearly 2,000 smaller dams in the Columbia-Snake River
basin. (See Figure 6.) Most of the big dams were built during the heyday of hydroelectric development 20 to 50 years
ago. Today, dams and reservoirs have blocked access to a
large portion of the fish's habitat.   Fragmentation of the
river system is so extreme that only 71 of its 1,996 kilometers run free. In the past, a young salmon's journey to the
ocean took two weeks; now it takes two months.[108]

The importance of looking at the cumulative impacts of
individual stresses is illustrated by the contribution of individual dams to salmon mortality. For example, if only 10
percent of the young salmon are killed at each dam they
encounter (and actual mortality rates are usually higher), by
the time the survivors pass all the dams on the journey
downriver, the cumulative mortality can approach 100 percent. A large proportion of the few that remain must make
an unnatural detour, and are hauled around the dams in
trucks and barges.[109]

The indirect obstacles to the salmon's journey—logging,
grazing, irrigation withdrawals, agricultural runoff, wetland
conversion—and rapidly expanding human populations
have all contributed to the extensive degradation of the
habitats of the Pacific Northwest over the last hundred
years. (See Figure 6.) Nearly 90 percent of its once extensive
primary forests have been lost to logging. Throughout the

FIGURE 6

Long-Term Stresses to the Columbia River Basin System, 1860–2000

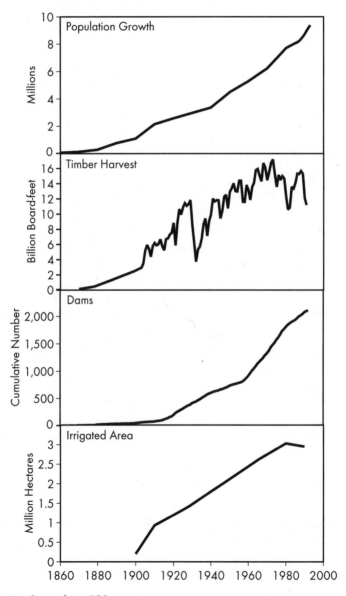

Source: See endnote 108

region, even undammed streams have lost much of their
salmon populations, in large measure because sedimentation and higher water temperatures caused by logging continue to render the streams unfit for fish for decades after
the operations have ceased.   Across the Canadian border,
the effects of overfishing, logging, and mining on
undammed major rivers such as the Fraser and the Skeena
have cut salmon populations to less than 20 percent of previously recorded levels. The loss of habitat has become so
severe throughout the region that the coho salmon is
extinct in 55 percent of its range and declining in 39 percent; it is considered not to be declining in just seven percent. The spring and summer chinook salmon are extinct in
63 percent of their range, and deteriorating in all but 6 percent. And not only migratory species such as the salmon
and sturgeon are imperiled. Many species that spend their
entire lives in one locale, so-called "resident species," are
also threatened with extinction by the same forces.[110]

Salmon populations are subject to some natural varia-
tion due to ocean currents and temperatures, amount of
rainfall, and so forth, but the cumulative pressures of
human assaults have severely impaired the species' ability to
cope with these natural forces. (See Figure 6.) Much of the
damage to the species' resilience is being done below the
species level, by the loss of particular "stocks."   Many
species, including salmon, are comprised of distinct groups
(or stocks) that have evolved unique adaptations to their
local environment. These fitness-enhancing characteristics
may govern time and place of spawning, migration routes,
and so forth. Fisheries biologists see these genetically different groups as the building blocks of conservation and rehabilitation.   Unfortunately, in the Pacific Northwest, they
have been sorely depleted. Of approximately one thousand
historic stocks, only one hundred are considered somewhat
healthy. Almost all of those are threatened by continued
habitat fragmentation and degradation, not to mention the
pressures of hatchery-bred fish.[111]

The decline of salmon and their habitats is causing serious conflicts between the fishing and timber industries, between conservation and utility groups, and between the United States and Canada. It has severely affected jobs and revenue throughout the region, and has led to fishing moratoriums on this once abundant species. It has also represented a loss to the Native American Indian tribes for whom salmon is an integral part of the culture and economy, and is guaranteed by treaty.[112]

There are some hopeful signs for the future, however. A number of detailed analyses have examined changes in dam operation, power generation, agriculture, logging, and irrigation that would be economically sound and benefit salmon recovery. Mitigating the negative impacts of current practices will improve the chances of recovery, but fundamental changes in land and water management and resource valuation are also required. A number of strategies have been put forth for rehabilitating or restoring habitat and species integrity and shifting to ecosystem management. The most recent and comprehensive one, from the National Research Council, has recommended institutional changes to provide for local flexibility and to allow management time scales and geographic scales that are large enough to be realistic. Virtually all plans have also called for protecting the genetic diversity of species as the cornerstone to salmon recovery. Another crucial step is supporting the shift that has already begun toward non-damaging economic uses of natural resources such as recreation and non-timber forest products that can provide a stable and sustainable economic base long into the future. (They already employ more people and bring in more revenue than logging.) Both the short- and long-term measures are needed or there may be no fish left to recover.[113]

Reconnecting the Fragments

Unfortunately, the plights of the Columbia, Mekong, Mississippi and Rhine rivers and the Great Lakes of North America and East Africa are not unique. Virtually every part of the globe has lost freshwater species and ecosystems (see Table 3). And none have escaped the cascade of unintended and unanticipated economic and social disruptions that follow the loss of healthy ecosystems—from more frequent and devastating floods and droughts, to the loss of clean and reliable water supplies, aquatic food resources, and livelihoods.[114]

Ironically the current piecemeal ways of managing freshwater systems often wreak havoc while trying to effect progress. Fisheries management, for instance, has long followed the agriculture production approach, "managing" each species individually, and using hatcheries and exotic fish stocking to prop up the fishing industry or placate recreational fishers. Overseas development programs frequently promote freshwater and coastal aquaculture instead of strengthening sound traditional fisheries practices that may hold the key to a sustainable future. Promoting hydropower as a path to prosperity at the expense of millions of peoples' sustainable livelihoods typically carves a path to deeper poverty. Adding fish passages to dams without addressing the habitat degradation caused by logging and grazing represents triage at best.

Until now, people have exploited and controlled freshwater resources in fragmented ways partly because the way they define ecosystems is itself fragmented. The tendency is to focus on only one element at a time—whether navigation, irrigation, power generation, sport fisheries, or even limited measures of water quality—without regard for the entire system. But a river does not stop at the water's edge; a healthy wetland is not simply a place with cattails and ducks.

The ways we view the goods and services provided by ecosystems, our understanding of our impacts, and our

TABLE 3

Freshwater Fish: Status and Threats

Area	Known freshwater fish species (number)	Extinct (percent)	Imperiled (percent)	Principal threats
Global	9000+	{ 20 combined }		
Amazon River	3000+	–	–	Habitat degradation Overharvest
Asia	1500+	–	–	Habitat degradation Competition for water Overharvest
North America	950	2	37	Habitat degradation Introduced Species
Mexico (arid lands)	200	8	60	Competition for water Pollution
Europe	193	–	42	Habitat degradation Pollution
South Africa	94	–	63	Habitat degradation Pollution Competition for water
Lake Victoria	350	57	43	Introduced Species
Costa Rica	127	–	9	Habitat degradation
Sri Lanka	65	–	28	
Iran	159	–	22	Habitat degradation Competition for water
Australia	188	–	35	Habitat degradation Competition for water

Source: See endnote 114

responses to the problems are also fragmented. We need to see a river or lake, along with its entire watershed and all its physical, chemical, and biological elements, as part of a complex, integrated system. Human inhabitants are also part of that system. Once we adopt a more comprehensive view of nature, we can learn to interact with ecosystems in ways that maintain their integrity, in what may be called ecosystem management. In an ecosystem-based approach, resources can be managed over large enough areas and long enough time scales to allow their species and ecological processes to remain intact while still allowing human activity. On a social level, involving all stakeholders in defining problems, setting priorities, and implementing solutions is essential.

In practical terms, adopting the principles of ecosystem management can begin by shifting perspectives towards recognizing all values of biodiversity and all those to whom biodiversity is of value. For example, an intact Nigerian floodplain supports tens of thousands of people through fishing, agriculture, fuelwood and fodder production, livestock, and tourism. Most households there engage in all of those activities and thus have a rather diverse and flexible strategy for sustaining themselves and adapting to changing conditions—it is their "insurance." The present economic value of the intact floodplain has been calculated at $1,990 per acre. When its current use was compared with the alternative of a water diversion plan it was calculated that water maintained in the floodplain was worth $45 per 1,000 cubic meters while the value of diverted water was only $0.04.[115]

But, not all of nature's "wealth" results from its current utility to humankind—it also has intrinsic values and it will be important to future generations. Nature's very existence is of great value to many people. The Bintuni Bay in Irian Jaya, Indonesia, has an estimated present value of around $1 billion (for direct uses such as fisheries and benefits from ecosystem services)—but it also has great cultural values to the local tribe. In this they are not unique—preserving nature's legacy and passing it on to future generations are fundamental values in virtually every culture.[116]

Because biodiversity has different values to different people and groups, adopting a broader view also requires broadening people's involvement in decisions. Many of the people who have the most to offer—and the most to lose— are usually excluded not only from management decisions but from the economic calculus as well. Along the Mekong, for example, the present and long-term value of the fisheries to the 52 million people who depend on them is seen by engineers and government officials as having little or no economic value when compared to the promise of quick infusions of revenue from large projects such as dams. Yet, decisions based on that equation will likely send the region into a social and ecological tailspin. Even when countries recognize the value of artisanal fisheries, they often attempt to make them more "modern" or "efficient," by replacing them with capital intensive, high-technology fishing industries. Such policies are not commercially viable, and they undermine both the resource base and the stability of the fishing communities.[117]

Many places are beginning to recognize the ecological and social benefits of community-based management and are upholding the rights of traditional resource users. Even in industrialized countries and altered ecosystems like the Columbia Basin some changes are underway. Balancing the demands of various industries and interests, pursuing less destructive uses of resources, and upholding the treaty rights of the area's original Indian inhabitants holds the most hope for restoring the once-bountiful salmon and ensuring the region's ecological and economic health.

Promoting open dialogue and public participation is an important part of sustainable management. At the local and regional level, watershed and basin management groups comprised of citizens, scientists, businesses, and governmental agencies are charting new courses for managing ecosystems and balancing various uses for the common good. Mechanisms such as the Global Convention on Biological Diversity, which encourage and support coordinated information sharing and action within and among countries, provide valuable opportunities for change. Cooperation and collaboration, with openness and transparency, are essential parts of ensuring our collective future.

Another essential step in ecosystem management is to halt further degradation and reduce the impact of human activities. In other words, tread lightly on the land and water. We can reduce the destructive impact of activities in the watershed such as logging and grazing and the disruption of riparian zones. We can protect "strongholds of aquatic biodiversity"—undegraded areas within now-troubled systems. The national wildlife refuges in the upper Mississippi Basin, for example, serve as havens for healthy populations that can repopulate troubled spots once they are restored. They also help maintain important ecosystem services. Further, hatcheries need not degrade the genetic base of ailing species; instead they can be used judiciously as a temporary measure to bolster them. And some simple techniques such as flushing the water from ships' ballasts while in the oceans rather than in freshwater harbors can greatly reduce the likelihood of accidental species introductions.

One way to minimize the harmful effects of energy development on people and nature has been suggested by Robert Goodland, a tropical ecologist at the World Bank. His method first considers energy efficiency measures and alternative sources of new energy. If hydropower is ultimately determined to be the best energy option, potential dam sites are then assessed by the area inundated and the number of people displaced per unit of energy. The idea is that a dam that inundates less area and displaces fewer people is preferable to one that covers more land and ousts more inhabitants per unit of energy. Using these criteria alone, it is clear that some of the recent and proposed dams are not worth the cost.[118]

Reducing the amount of chemicals and wastes we generate and add to the waters has been one of the greatest environmental successes of the last 20 years. The Clean

Water Act in the U.S. and the Rhine Action Plan in Europe have brought about significant improvements in water quality by reducing pollution from industrial and municipal waste. Better sewage treatment and changes in industrial processes have cut the amount of heavy metals entering the Rhine by 90 percent since the 1970s. On both continents, removing lead from gasoline and phosphates from detergents has substantially decreased the amount of those harmful substances in the water at little cost to consumers or industry. Reducing the pollutants that enter from diffuse sources such as runoff from farms and homes remains a challenge.[119]

Once human stresses are removed, and natural processes supported, a system can often become self-sustaining again.

Another necessary step towards ecosystem management is to expand and integrate resource management to encompass larger areas and longer time scales in order to maintain all of a system's processes and components. Many of today's problems resulted from earlier short-sighted attempts to solve other problems (such as introducing Nile perch to Lake Victoria several decades ago). Both the concept of sustainability and the nature of ecosystems require that decisions affecting them look not only beyond the next few years but beyond the next few decades as well. Regional efforts like those on the Rhine River and around Lake Victoria that set long-term goals and include all countries are starting to reflect this reality. The need for a broad geographic view—that is managing at the ecosystem level—is especially apparent in aquatic systems because the impact of every activity within the watershed (and many beyond) is ultimately revealed in the water.

Much fragmented management arises because institutional responsibilities are fragmented. Often various agencies have missions that work at cross purposes—jurisdictions and mandates may contradict each other as well as sound long-range management. In the Mississippi Basin, for example, numerous agencies operating at the federal, state and local levels are responsible for various aspects of navigation, soil conservation, water resources management, coastal management, wetlands management, wildlife management, flood control, disaster assistance, building permits, etc. The 1993 Mississippi flood and its aftermath prompted a revision of the nation's program for floodplain management to improve coordination of all these functions with two broad goals in mind: reducing the loss of life and damage done by floods, and preserving and restoring the natural resources and functions of floodplains.[120]

One way institutions might take broader and longer views would be to shift from narrow extractive approaches to more broadly managing for the full array of ecosystem goods and services. For example, rather than propping up the stocks of a few ailing commercial species through hatcheries, they might help rivers to return to a self-sustaining condition. These rivers could be productive over the long term, without constant human intervention. They could also provide the array of services that we depend on.

In much of the world, the landscape has already been highly altered. As we try to restore the health of modified freshwater systems it is clearly not enough to just focus on individual species or isolated habitats. It is critical to maintain and restore the ecosystems' processes. Often a first step involves actions such as mitigating pollution, controlling exotics, and improving the flow and quality of water. In many places, rather than attempting to restore a system to its former pristine state, a more realistic goal may be to rehabilitate or revive some ecosystem functions and encourage nature's own regenerative processes. Once human pressures and stresses are removed, and natural processes are supported (such as ensuring that enough water flows through the system, at the right time of year), a system can often achieve a self-sustaining state. It may not be quick, but it is more likely to be long lasting.[121]

Part of restoration may include removing aging or inefficient infrastructure at the end of its life span, or eliminating perverse economic subsidies. For example, an economic analysis of the slated removal of two aging dams on the Elwha River in Washington state found that the $100 million price tag is more than justified by the estimated $3 to $6 billion per year in non-market benefits (such as recreation, existence and bequest values) that river and wild salmon restoration would bring. Along the Missouri River, by removing the subsidies and structures that support uneconomical navigation, the natural flow and course of the river can begin to be restored. By restoring the dynamic forces that created and maintained the river, the river can better support its flora and fauna and the human uses of the river that have higher value and are less destructive. Nature wins, the economy wins, and people win.[122]

Restoration and rehabilitation need not be highly disruptive to existing uses and users of the landscape. When flood-damaged levees are rebuilt, for example, placing them further back from the river's edge can allow more of the natural floodplain to function and thus reduce future flood damage to human settlements and economic activities. An extensive study by the U.S. National Research Council recommended rehabilitation and restoration of aquatic ecosystems to solve water quality, wildlife, and flooding problems at minimal cost and disruption. Restoring 50 percent of the wetlands lost in the United States would affect less than 3 percent of the nation's agricultural, forest, or urban land. The costs of such restoration would be relatively small and the benefits large. Along the Missouri and upper Mississippi rivers, buying land from farmers willing to sell can return acreage to the river system and thus reduce the costs and impact of future floods. While the cost of rehabilitation or restoration may seem high, it is far less than the price of continued mismanagement.[123]

Restoration and rehabilitation alone, however, will not be enough. A principal goal now and in the future should be to shift from reactive to preventive management.[12] Concerted cooperation and action will be required from the local stream bank to international diplomatic halls. Rivers like the Mekong and the Amazon offer chances to avoid the costly and irreversible mistakes made elsewhere. We already know the heavy price that regulated rivers, dam construction, water diversion and pollution, alien species, fisheries management, and habitat degradation and fragmentation can exact from a region. The time has come to act on a corollary principle: over the long term, keeping naturally functioning ecosystems healthy will offer the greatest number of benefits for the greatest number of people. The health of the planet's circulatory system and our own chances for a healthy future depend on it.

Notes

1. Arthur E. Bogan, "Freshwater Bivalve Extinctions (Mollusca: Unionidae): A Search for Causes," *American Zoology* 33:599-609, 1993.

2. Ibid.

3. Paul W. Parmalee and Walter E. Klippel, "Feshwater Mussels as a Prehistoric Food Resource," *American Antiquity*, July 1994; Paul W. Parmalee, Walter E. Klippel, and Arthur E. Bogan, "Aboriginal and Modern Freshwater Mussel Assemblages (Pelecypoda: Unionidae) From the Chickamauga Reservoir, Tennessee," *Brimleyana*, December 1982; Larry Master, "Aquatic Animals: Endangerment Alert," *The Nature Conservancy*, March/April 1991; James D. Williams et al., "Conservation Status of Freshwater Mussels of the United Sates and Canada," *Fisheries*, September 1993; The Nature Conservancy, *Priorities for Conservation: 1996 Annual Report Card for U.S. Plant and Animal Species* (Arlington, Va.: 1996).

4. Bogan, op. cit. note 1; mussel/herring example from S.L.H. Fuller, "Historical and Current Distributions of Freshwater Mussels (Mollusca: Bivalvia: Uniondae) in the Upper Mississippi River," in J. Rasmussen, ed., *Proceedings of the UMRCC Symposium on Upper Mississippi River Bivalve Mollusks* (Rock Island, Ill.: Upper Mississippi River Conservation Committee, 1980), cited in Richard E. Sparks, "Need for Ecosystem Management of Large Rivers and Their Floodplains," *Bioscience*, March 1995. Paul W. Parmalee and Mark H. Huges, "Freshwater Mussels (Mollusca: Pelecypoda: Unionidae) of Tellico Lake: Twelve Years After Impoundment of the Little Tennessee River," *Annals of the Carnegie Museum*, 25 February 1995.

5. Water resource estimates from Alan P. Covich, "Water and Ecosystems," in Peter H. Gleick, ed., *Water in Crisis: A Guide to the World's Fresh Water Resources* (New York: Oxford University Press, 1993); Peter B. Moyle and Robert A. Leidy, "Loss of Biodiversity in Aquatic Ecosystems: Evidence from Fish Faunas," in P.L. Fiedler and S.K. Jain, eds., *Conservation Biology: The Theory and Practice of Nature Conservation, Preservation, and Management* (New York: Chapman and Hall, 1992); The Nature Conservancy 1996, op. cit. note 3; Jack E. Williams et al., "Fishes of North America Endangered, Threatened or of Special Concern: 1989," *Fisheries*, November-December 1989; Robert R. Miller, James D. Williams, and Jack E. Williams, "Extinctions of North American Fishes During the Past Century," *Fisheries*, November-December 1989. **Figure 1** from Covich. **Figure 2** from The Nature Conservancy 1996, op. cit. note 3.

6. Extinction rates from Stuart L. Pimm et al., "The Future of Biodiversity," *Science*, July 21, 1995; "biodiversity deficit" from Jonathan Coddington, speech at "The Living Planet in Crisis: Biodiversity Science and Policy," American Museum of Natural History, New York, N.Y., March

9-10, 1995. Species known from Edward O. Wilson, *The Diversity of Life* (New York: W.W. Norton & Company, 1992); J. David Allan and Alexander S. Flecker, "Biodiversity Conservation in Running Waters," *Bioscience*, January 1993.

7. Allan and Flecker, op. cit. note 6.

8. Malaysia example from David Pearce and Dominic Moran, *The Economic Value of Biodiversity*, (London: Earthscan, 1994); Florida example from J.M. Hefner et al., *Southeast Wetlands: Status and Trends, Mid-1970s to Mid-1980s*, (Atlanta, Georgia: U.S. Department of the Interior [DOI], Fish and Wildlife Service [FWS], 1994); T.E. Dahl, *Wetland Losses in the United States 1780's to 1980's* (Washington, D.C.: FWS, 1990).

9. United Nations Food and Agriculture Organization (FAO), *State of World Fisheries and Aquaculture* (Rome: 1995); Great Lakes fishery from Robert M. Hughes and Reed F. Noss, "Biological Diversity and Biological Integrity: Current Concerns for Lakes and Streams," *Fisheries*, May-June 1992.

10. Barry L. Johnson, William B. Richardson, and Teresa J. Naimo, "Past, Present and Future Concepts in Large River Ecology," *Bioscience*, March 1995; J.V. Ward and J.A. Stanford, "Riverine Ecosystems: The Influence of Man on Catchment Dynamics and Fish Ecology," in D.P. Dodge, ed., *Proceedings of the International Large River Symposium*, Canadian Special Publication of Fisheries and Aquatic Sciences 106, (Ottawa: Department of Fisheries and Oceans, 1989).

11. Peter B. Bayley, "Understanding Large River-Floodplain Ecosystems," *Bioscience*, March 1995; Ward and Stanford, op. cit. note 10.

12. Mats Dynesius and Christer Nilsson, "Fragmentation and Flow Regulation of River Systems in the Northern Third of the World," *Science*, November 4, 1994; A.C. Benke, "A Perspective on America's Vanishing Streams," *Journal of the North American Benthological Society*, 1990, cited in David S. Wilcove and Michael J. Bean, eds., *The Big Kill: Declining Biodiversity in America's Lakes and Rivers* (Washington, D.C.: Environmental Defense Fund (EDF), 1994); G.E. Petts cited in Covich, op. cit. note 5.

13. **Table 1** from Robert J. Naiman, et al., eds., *The Freshwater Imperative: A Research Agenda*, (Washington, DC.: Island Press, 1995) (as adapted from M.I. L'vovich and G.F. White, "Use and Transformation of Water Systems," in B.L. Turner II et al., eds., *The Earth as Transformed by Human Action: Global; and Regional Changes in the Biosphere Over the Past 300 Years*, (Cambridge: Cambridge University Press, 1990) except: dam data from World Resources Institute, *World Resources 1992-93*, (New York: Oxford University Press, 1992).

14. Antonin Lelek, "The Rhine River and Some of Its Tributaries Under

Human Impact in the Last Two Centuries," in Dodge, ed., op. cit. note 10; drinking water information from Marlise Simons, "Salmon Does Not Mean the Rhine's Water is Safe to Drink," *New York Times*, May 25, 1995.

15. 1995 flood data from Haig Simonian, "Flood of Tears on the Rhine," *Financial Times*, February 8, 1995; historic flood data from "Dyke Disaster," *Down to Earth*, March 15, 1995.

16. Rhine ecology data from Lelek, op. cit. note 14.

17. Fred Pearce, "Greenprint for Rescuing the Rhine," *New Scientist*, June 26, 1995; effects of flood control devices from Edward Goldsmith and Nicholas Hildyard, *The Social and Environmental Effects of Large Dams, Volume One: Overview* (Camelford, Cornwall, U.K.: Wadebridge Ecological Centre, 1984).

18. Lelek, op. cit. note 14; Pearce, op. cit. note 17.

19. Simonian, op. cit. note 15; only two of the twenty designated areas have been completed; Pearce, op. cit. note 17.

20. Gerald E. Galloway, "The Mississippi Basin Flood of 1993," prepared for Workshop on Reducing the Vulnerability of River Basin Energy, Agriculture and Transportation Systems to Floods, Foz do Iguacu, Brazil, November 29, 1995; levee damage from Mary Fran Myers and Gilbert F. White, "The Challenge of the Mississippi Flood," *Environment*, December 1993.

21. FWS, *Figures on Wetlands Lost in Mississippi Basin Prepared for Post Flood Recovery and the Restoration of Mississippi Basin Floodplains Including Riparian Habitat and Wetlands* (St. Louis, Mo.: Association of State Wetland Managers (ASWM), 1993), cited in Wilcove and Bean, eds., op. cit. note 12.

22. Evolution of Mississippi River management and early estimates of 1993 flood costs from Myers and White, op. cit. note 20; artificial channel length from Jeff Hecht, "The Incredible Shrinking Mississippi Delta," *New Scientist*, April 14, 1990; flood heights from L.B. Leopold, "Flood Hydrology and the Floodplain," in Gilbert F. White and Mary Fran Myers, eds., *Water Resources Update: Coping with the Flood: The Next Phase* (Carbondale, Ill.: University Council on Water Resources, 1994), cited in Sparks, op. cit. note 4; historic flood costs from William Stevens, "The High Costs of Denying Rivers Their Floodplains," *New York Times*, July 20, 1993; 1993 costs from Galloway, op. cit. note 20.

23. Deborah Moore, "What Can We Learn From the Experience of the Mississippi?" Environmental Defense Fund, San Francisco, Calif., September 7, 1994.

24. Missouri River Coalition (MRC), "Comments on the Missouri River Master Water Control Manual Review and Update Draft Environmental

Impact Assessment," March 1, 1995.

25. Myers and White, op. cit. note 20; flood payments in Robert S. Devine, "The Trouble With Dams," *Atlantic Monthly*, August 1995.

26. MRC, op. cit. note 24.

27. For summary of Task Force recommendations, see Myers and White, op. cit. note 20; Galloway, op. cit. note 20; National Research Council (NRC), *Restoration of Aquatic Ecosystems: Science, Technology, and Public Policy* (Washington, D.C.: National Academy Press, 1992).

28. For wetland losses by state see Dahl, op. cit. note 8; Calvin R. Fremling et al., "Mississippi River Fisheries: A Case History," in Dodge, ed., op. cit. note 10, and Joseph H. Wlosinski et al., "Habitat Changes in Upper Mississippi River Floodplain" in E.T. LaRoe et al., eds., *Our Living Resources: a Report to the Nation on the Distribution, Abundance, and Health of U.S. Plants, Animals, and Ecosystems* (Washington, D.C.: DOI, National Biological Service, 1995); MRC, op. cit. note 24; L.W. Hesse and G.E. Mestl, "The Status of Nebraska Fishes in the Missouri River. 1. Paddlefish (Polyodontidae: Polyodon Spathula)," *Transactions of the Nebraska Academy of Science*, cited in MRC, op. cit. note 24.

29. R.L. Welcomme, "River Fisheries," Fisheries Technical Paper 262, FAO, Rome, 1985, cited in Sparks, op. cit. note 4; National Marine Fisheries Service, *Fisheries of the United States, 1990* (Washington, D.C.: U.S. Government Printing Office (GPO), 1991), cited in Hefner, op. cit. note 8.

30. For more on declines, see James Wiener et al., "Biota of the Upper Mississippi River Ecosystem" in LaRoe, et al., eds., op. cit. note 28; Sparks, op. cit. note 4; for more on the Yangtze River, see Audrey Topping, "Ecological Roulette: Damming the Yangtze," *Foreign Affairs*, September/October 1995.

31. Raphael Heath, "Hell's Highway," *New Scientist*, June 3, 1995.

32. "Inter-Governmental Committee on Hidrovia Initiates Plans for Paraguay River Alterations," *Minutes of CIH Meeting, November 9, 1994* in Hidrovia Campaign, Dossier 1, March 1, 1995; Fundaçao Centro Brasileiro de Referencia e Apoio Cultural (CEBRAC) and World Wildlife Fund (WWF), "Parana-Paraguay Waterway: Who Pays the Bill?" Executive Summary, CEBRAC, Brasilia, September 1994.

33. Species numbers from Covich, op. cit. note 5; CEBRAC and WWF, op. cit. note 32; Heath, op. cit. note 31.

34. Marcelo Jardim, President, Inter-Governmental Committee on Hidrovia, briefing at the World Wildlife Fund, Washington, D.C., September 21, 1995.

35. Philip M. Fearnside quoted in Allan and Flecker, op. cit. note 6; World Resources Institute, *World Resources 1992-93* (New York: Oxford University Press, 1992); Devine, op. cit. note 25.

36. "The Beautiful and the Dammed," *Economist*, March 28, 1992.

37. Displacement figures from Robert Goodland, "Environmental Sustainability Needs Renewable Energy: The Extent to Which Big Hydro is Part of the Transition," presented to The Mekong: International Technical Workshops for Sustainable Development Through Cooperation, Washington, D.C., November 28-December 2, 1995; Three Gorges displacement from Human Rights Watch (HRW), "The Three Gorges Dam in China," *Human Rights Watch/Asia*, February 1995.

38. Water and Sanitation Health Project (WSHP), *Senegal River Basin Health Master Plan Study*, cited in Aleta Brown, "Dying for Dams: Decade of River Development Leads to Health Disaster in Senegal," *World Rivers Review*, November 1995; R.L. Welcomme, "Relationships Between Fisheries and the Integrity of River Systems," *Regulated Rivers: Research & Management*, 11:121-136, 1995; M.M. Horowitz, "The Management of an African River Basin: Alternative Scenarios for Environmentally Sustainable Economic Development and Poverty Alleviation," from *Proceedings of the International UNESCO Symposium: Water Resources Planning in a Changing World*, Karlsruhe, Germany, June 28-30, 1994.

39. Anne E. Platt, "Confronting Infectious Diseases," in Lester R. Brown et al., *State of the World 1996* (New York: W.W. Norton, 1996); WSHP, op. cit. note 38.

40. Moyle and Leidy, op. cit. note 5.

41. Allan and Flecker, op. cit. note 6; Naiman, et al., eds., op. cit. note 13; Covich, op. cit. note 5; Welcomme, op. cit. note 38.

42. Goodland, op. cit. note 37; Philip M. Fearnside, "Hydroelectric Dams in the Brazilian Amazon as Sources of 'Greenhouse' Gases," *Environment Conservation*, Spring 1995.

43. HRW, op. cit. note 37; Topping, op. cit. note 30; Dai Qing, *Yangtze! Yangtze!* (London: Earthscan, 1989).

44. Post-dam wild salmon information from FWS, "Pacific Salmon Management," briefing document, Region 1 Fisheries, Portland, Ore., 1991, cited in James B. Petit, "Solid Faith in Small Acts," *Ilahee*, Winter 1994; M.C. James, "Report of the Division of Commercial Fishing," *Transactions of the American Fisheries Society*, Washington, D.C., 1938; Bureau of Reclamation quote from N. Tangwisutijit, "Reclaiming Respect for Rivers: A Conversation with Dan Beard," *World Rivers Review*, Fourth Quarter 1994.

45. Richard J. Grant, "Go With the Flow," *Worldlink*, July/August 1995.

46. William Barnes, "Dash to Dam the Mekong Raises Ecology Fears," *Financial Times*, December 14, 1994.

47. Yuan Shu, "Nations Find Unity in Taming the Mekong," *World Paper*, November 1994; fish migrations from Mark T. Hill and Susan A. Hill, "Summary of Fisheries Resources and Projects in the Mekong River," presented at The Mekong: International Technical Workshops for Sustainable Development Through Cooperation, Washington, D.C., November 28-December 2, 1995; spawning fish estimate from Barnes, op. cit. note 46; population data from "New Mekong River Basin Agreement Will Spur Hydro Development, Groups Charge," *International Environment Reporter*, April 19, 1995; Ian G. Baird, "Community Management of Mekong River Resources in Laos," *Naga: The ICLARM Quarterly*, (International Center for Living Aquatic Resources Management (ICLARM), October 1994.

48. Mok Moreth, "Environmental Concern Facing Cambodia," presented at The Mekong: International Technical Workshops for Sustainable Development Through Cooperation, Washington, D.C., November 28-December 2, 1995; Denise Heywood, "Reversal of Fortune," *Geographical*, November 1994.

49. Moreth, op. cit. note 48.

50. "New Mekong River Basin Agreement," op. cit. note 47.

51. Barnes, op. cit. note 46.

52. Electricity demands from Shu, op. cit. note 47; "Statement on Cooperation for the Sustainable Development of the Mekong River Basin," Thai non-governmental organizations, Econet posting, April 4, 1995.

53. "Mekong Politics: 'New Era', Same Old Plans," *Watershed: People's Forum on Ecology*, June 1995; Grainne Ryder, "Overview of Regional Plans," *World Rivers Review*, Fourth Quarter 1994.

54. Editorial, *The Nation* March 12, 1995; "Lao Dam Deal at Risk, Thai Electricity Demand Falling," *Watershed: People's Forum on Ecology*, November 1995-February 1996.

55. Dave Hubbel, "Thailand's Pak Mun Dam: A Case Study," *World Rivers Review*, Fourth Quarter 1994; compensation information from "Community Voices: Speaking Out on the Pak Mun Dam" *Watershed: People's Forum on Ecology*, June 1995.

56. Reports of logging from William Robichaud, "Matters and Species of Conservation Concern in the Mekong Watershed," presented at The Mekong: International Technical Workshops for Sustainable Development

Through Cooperation, Washington, D.C., November 28-December 2, 1995; forest cover estimates from Moreth, op. cit. note 48; timber concessions from Bernhard O'Callaghan, Asian Wetlands Bureau, private communication, November 30, 1995; "Cambodia's Forests Sold to Asian Timber Barons," *Watershed: People's Forum on Ecology*, November 1995-February 1996.

57. Barnes, op. cit. note 46.

58. Sandra Postel, "Water in Agriculture," in Gleick, ed., op. cit. note 5.

59. Sandra Postel, "Rivers Drying Up," *World Watch*, May/June 1995; Moyle and Leidy, op. cit. note 5.

60. Postel, op. cit. note 59; "UN to Assess Aral Sea Shrinkage," *Financial Times*, September 19, 1995; Judith Perera, "A Sea Turns to Dust," *New Scientist*, October 23, 1993.

61. NRC, op. cit. note 27; Moody et al. cited in Dynesius and Nilsson, op. cit. note 12; Postel, op. cit. note 59.

62. NRC, op. cit. note 27. **Figure 3** data from Igor A. Shiklomanov, "World fresh water resources," in Peter H. Gleick, ed., *Water in Crisis: A Guide to the World's Fresh Water Resources* (New York: Oxford University Press, 1993).

63. Ricki Lewis, "Can Salmon Make a Comeback," *Bioscience*, January 1991; Covich, op. cit. note 5.

64. Population data from James L. Tyson, "Delicate Ecosystem, Great Lakes Weighs the Economic Demands of Heavy Industry Manufacturing With the Environment's Needs," *Christian Science Monitor*, March 14, 1994; industrial and agricultural activity from Steve Thorp and David R. Allardice, *A Changing Great Lakes Economy: Economic and Environmental Linkages*, Working Paper, State of the Lakes Ecosystem Conference, (Ann Arbor, Michigan: Great Lakes Commission, October 1994); Nature Conservancy, Great Lakes Program, *Conservation of Biological Diversity in the Great Lakes Ecosystem: Issues and Opportunities* (Chicago: 1994); wetlands loss from Ronald E. Erickson, "The National Wetlands Inventory in the Great Lakes Basin of the United States," in *Wetlands of the Great Lakes: Protection and Restoration Policies; Status of the Science—Proceedings of an International Symposium* (New York: Association of State Wetland Managers, 1994); water quality from United States Environmental Protection Agency (EPA), *The Quality of Our Nation's Water: 1992* (Washington, D.C.: 1994).

65. Historical overview of pollution from George R. Francis and Henry A. Reiger, "Barriers and Bridges to the Restoration of the Great Lakes Basin Ecosystem," in Lance H. Gunderson et al., eds., *Barriers & Bridges to the Renewal of Ecosystems and Institutions* (New York: Columbia University Press, 1995); Richard Sparks, "The Illinois River-Floodplain Ecosystem," in NRC,

op. cit. note 27.

66. Basin outflow estimate from United States General Accounting Office, *Pesticides: Issues Concerning Pesticides Used in the Great Lakes Watershed* (Washington, D.C.: June 1993); Tyson, op. cit. note 64; airborne pollutant load from Michigan Department of Natural Resources (MDNR), *State of the Great Lakes: 1993 Annual Report* (Lansing, Michigan: Office of the Great Lakes, 1993); distant origins of some airborne pollutants from Barry Commoner, reported in *International Environment Reporter*, May 31, 1995; Theodora E. Colborn et al., *Great Lakes Great Legacy?* (Washington, D.C.: Conservation Foundation, 1990); basin size data from Nature Conservancy, op. cit. note 64.

67. Fish consumption advisories from EPA, op. cit. note 64; for complete database, see EPA, Office of Water, *National Listing of Fish Consumption Advisories Database*, July 1995; estimates of chemicals entering the system from Tyson, op. cit. note 64; number of chemicals monitored from Francis and Reiger, op. cit. note 65.

68. Theodora E. Colborn, "Global Implications of Great Lakes Wildlife Research," *International Environmental Affairs*, Winter 1991; Theodora E. Colborn and Coralie Clement, eds., *Chemically-Induced Alterations in Sexual and Functional Development: The Wildlife/Human Connection* (Princeton, N.J.: Princeton Scientific Publishing, 1992); Tyson, op. cit. note 64.

69. Colborn, op. cit. note 68; Colborn and Clement, eds., op. cit. note 68; Robert J. Hesselberg and John E. Gannon, "Contaminant Trends in Great Lakes Fish," in LaRoe et al., eds., op. cit. note 28; E. Carlsen et al., "Evidence for Decreasing Quality of Semen During Past 50 Years," *British Medical Journal*, Vol. 305, 1992, pp. 609-613, cited in Sue Dibb, "Swimming in a Sea of Oestrogens: Chemical Hormone Disrupters," *Ecologist*, January/February 1995.

70. For progress under Great Lakes Water Quality Agreement, see Francis and Reiger, op. cit. note 65; Hesselberg and Gannon, op. cit. note 69; EPA, *A Phase I Inventory of Current EPA Efforts to Protect Ecosystems* (Washington, D.C.: 1995).

71. FAO, *Review of the State of the World Fishery Resources: Inland Capture Fisheries*, Fisheries Circular No. 885 (Rome: 1995). **Table 2** data for Asia, Europe, South America from Welcomme, op. cit. note 38, and FAO; West Africa from FAO; Lake Victoria from Lowe-McConnell, op. cit. note 94; Missouri River from Missouri River Coalition, op. cit. note 24 and Welcomme, op. cit. note 38; Illinois River from Richard E. Sparks, "The Illinois River-Floodplain Ecosystem," in NRC op. cit. note 27; North American Great Lakes from Hughes and Noss, op. cit. note 9; Aral Sea from op. cit. note 60.

72. Welcomme, op. cit. note 38; Reed F. Noss and Allen Y. Cooperrider,

Saving Nature's Legacy: Protecting and Restoring Biodiversity (Washington, D.C.: Island Press, 1994).

73. ICLARM, Consultative Group on International Agricultural Research (CGIAR), *From Hunting to Farming Fish* (World Bank: Washington, D.C., 1995).

74. ICLARM, op. cit. note 73; FAO, op. cit. note 71.

75. FAO, op. cit. note 9.

76. Ibid.

77. Ibid.; Nigel J.H. Smith, *Man, Fishes and the Amazon* (New York: Columbia University Press, 1981).

78. Forest loss from Michael Goulding, "Flooded Forests of the Amazon," *Scientific American*, March 1993; river statistics from Igor A. Shiklomanov, "World Fresh Water Resources," in Gleick, ed., op. cit. note 5; flooded forest harvest from Janet M. Chernela, "Tukanoan Fishing," *National Geographic Research & Exploration*, vol. 10, no. 4, 1994; fish catch and consumption from Peter B. Bayley and Miguel Petrere, Jr., "Amazon Fisheries: Assessment Methods, Current Status and Management Options," in Dodge, ed., op. cit. note 10; tambaqui from Eliot Marshall, "Homely Fish Draws Attention," *Science*, February 10, 1995.

79. National Research Council, *Upstream: Salmon and Society inthe Pacific Northwest* (Washington, D.C.: National Academy Press, 1995); Willa Nehlsen, Jack E. Williams, and James A. Lichatowich, "Pacific Salmon at the Crossroads: Stocks at Risk From California, Oregon, Idaho and Washington," *Fisheries*, March-April 1991; Jessica Maxwell, "Swimming With Salmon," *Natural History*, September 1995; The Conservation Fund and National Fish and Wildlife Foundation, *Report of the National Fish Hatchery Review Panel*, Washington, D.C., December 30, 1994; Pacific Rivers Council (PRC), *Coastal Salmon and Communities at Risk: The Principles of Coastal Salmon Recovery* (Eugene, Oregon: July 1995); Jack K. Sterne, Jr., "Supplementation of Wild Salmon Stocks: A Cure for the Hatchery Problem or More Problem Hatcheries," *Coastal Management*, 23:123-152 1995.

80. FAO, op. cit. note 9.

81. Ibid.; FAO, op. cit. note 71; Chen Chunmei, "Aquaculture Meets Demand for Seafood," *China Daily*, December 26, 1995.

82. FAO, op. cit. note 71.

83. Malcolm C. Beveridge, Lindsay G. Ross, and Liam A. Kelly, "Aquaculture and Biodiversity," *Ambio*, December 1994.

84. Ibid.; FAO, *Aquaculture Production Statistics 1984-1993*, Fisheries Circular No. 815, Revision 7, Rome, 1995.

85. Beveridge, Ross and Kelly, op. cit. note 83.

86. Miller, Williams, and Williams, op. cit. note 5.

87. Melanie Stiassney, "An Overview of Freshwater Biodiversity: With Some Lessons From African Fishes," *Fisheries*, in press.

88. Charles Boydstun, Pam Fuller, and James D. Williams, "Nonindigenous Fish," in LaRoe et al., op. cit. note 28; number of species established from Billy Goodman, "Keeping Anglers Happy Has a Price," *Bioscience*, May 1991; tropical introductions from Beveridge, Ross, and Kelly, op. cit. note 83.

89. Covich, op. cit. note 5.

90. Alliance for the Chesapeake Bay, "Species Invasions Around the World That Have Brought Havoc," *Bay Journal*, April 1995; John T. Carleton, quoted in "Species Invasions Around the World Have Brought Havoc," *Bay Journal* April 1995.

91. Historic fish catches from Hughes and Noss, op. cit. note 9.

92. Number of exotics from EPA, op. cit. note 70; post-seaway estimates from *Great Lakes Panel on Aquatic Nuisance Species: Annual Report* (Ann Arbor, Mich.: Great Lakes Commission, March 1995); lamprey impact on lake trout catch and initiation of Great Lakes Fishery Commission from Colborn et al., op. cit. note 66; lamprey control information from MDNR, op. cit. note 66.

93. Tom Kenworthy, "Zebra Mussels May Threaten California Irrigation System," *Washington Post*, August 22, 1995; Great Lakes Panel, op. cit. note 92; Michael L. Ludyanskiy et al., "Impact of the Zebra Mussel, a Bivalve Invader," *Bioscience*, September 1993.

94. Rosemary Lowe-McConnell, "Fish Faunas of the African Great Lakes: Origins, Diversity and Vulnerability," *Conservation Biology*, September 1993; E. Barton Worthington and Rosemary Lowe-McConnell, "African Lakes Reviewed: Creation and Destruction of Biodiversity," *Environmental Conservation*, Autumn 1994; lake size from Igor A. Shiklomanov, "World Fresh Water Resources," in Gleick, ed., op. cit. note 5.

95. Les Kaufman, "Catastrophic Change in Species-Rich Freshwater Ecosystems: The Lessons of Lake Victoria," *Bioscience*, December 1992; endemic fish loss estimates from Lowe-McConnell, op. cit. note 94.

96. Journalists Environmental Association of Tanzania (JEAT) and Panos Institute, *Current State of the Lake Report* (London: Panos Institute, May

1994).

97. Lowe-McConnell, op. cit. note 94; Kaufman, op. cit. note 95; commercial catch data from "Fishing Industry Devours Itself," *Panoscope*, July 1994. Figure 4 from Kenya Marine and Fisheries Research Institute in Les Kaufman, op. cit. note 95.

98. Kaufman, op. cit. note 95.

99. "Battling the Killer Weed," *Panoscope*, July 1994.

100. "Lake Victoria's Sea of Troubles," *Panoscope*, July 1994; JEAT and Panos Insitute, op. cit. note 96; "Tide of Horror From Rwandan War," *Panoscope*, July 1994.

101. Les Kaufman, private communication, October 11, 1995.

102. Miller, Williams, and Williams, op. cit. note 5.

103. Naiman et al., eds., op. cit. note 13.

104. Dynesius and Nilsson, op. cit. note 12.

105. Mary F. Wilson and Karl C. Halupka, "Anadromous Fish as Keystone Species in Vertebrate Communities," *Conservation Biology*, June 1995.

106. Historic harvests from Oregon Department of Fish and Wildlife (ODFW) and Washington Department of Fish and Wildlife (WDFW), *Status Report Columbia River Fish Runs and Fisheries, 1938–1994*, (Portland, Oregon: ODFW and WDFW, 1995); additional data on historic populations and harvest from Carolyn Alkire, *The Living Landscape, Vol. 1: Wild Salmon as Natural Capital*, (Washington, D.C.: Wilderness Society, August 1993); NRC, op. cit. note 79; John C. Ryan, *State of the Northwest*, Northwest Environment Watch Report 1 (Seattle, Washington: 1994); Sterne, op. cit. note 79; Tom Kenworthy, "Agency Outlines Salmon Protection Plan," *Washington Post*, March 21, 1995. Figure 5 from Oregon Department of Fish and Wildlife (ODFW) and Washington Department of Fish and Wildlife (WDFW), *Status Report Columbia River Fish Runs and Fisheries, 1938–1994*, (Portland, Oregon: ODFW and WDFW, 1995).

107. Adam Diamant and Zach Wiley, *Water for Salmon: An Economic Analysis of Salmon Recovery Alternatives in the Lower Snake and Columbia Rivers*, prepared for the Northwest Power Planning Council (New York: Environmental Defense Fund, April 1995).

108. Dams in basin from Diamant and Wiley, op. cit. note 107; NRC, op. cit. note 79; extent of river impounds from Wilcove and Bean, op. cit. note 12; salmon migration time from Ryan, op. cit. note 106. Figure 6 population growth from U.S. Bureau of the Census; dams and timber harvest

from National Research Council, op. cit. note 79; irrigated area from *Columbia River System Operation Review: Final Environment Impact Statement (Appendix F Irrigation, Municipal and Industrial/Water Supply)* (DOE/EIS-0170), (U.S. Department of Energy, November 1995).

109. NRC, op. cit. note 79.

110. Forest lost from Douglas E. Booth, "Estimating Prelogging Old-Growth in the Pacific Northwest," *Journal of Forestry*, October 1991; impact of logging from Christopher A. Frissell, *A New Strategy for Watershed Restoration and Recovery of Pacific Salmon in the Pacific Northwest* (Corvallis, Oregon: PRC, 1993); coho listing from "Coho Salmon Proposed as 'Threatened Species'," *New York Times*, July 21, 1995; Canadian salmon data from T.G. Northcote and D.Y. Atagi, "Pacific Salmon Abundance Trends in the Fraser River Watershed Compared With Other British Columbia Systems," in Deanna J. Stouder, Peter A. Bisson, and Robert J. Naiman, eds., *Pacific Salmon and their Ecosystems: Status and Future Options* (New York: Chapman & Hall, in press); range extinction data from Wilderness Society, *The Living Landscape, Vol. 2: Pacific Salmon and Federal Lands* (Washington, D.C.: October 1993); non-migratory species status from M. G. Henjum et al., *Interim Protection for Late-Successional Forests, Fisheries, and Watersheds: National Forests East of the Cascade Crest, Oregon and Washington* (Bethesda, Md.: Wildlife Society, 1994).

111. Natural variations from James A. Lichatowich and Lars E. Mobrand, *Analysis of Chinook Salmon in the Columbia River From an Ecosystem Perspective*, research report prepared for U.S. Department of Energy (Vashon Island, Wash.: Mobrand Biometrics, January 1995); NRC, op. cit. note 79; Nehlsen, Williams, and Lichatowich, op. cit. note 79; Charles W. Huntington, Willa Nehlsen and Jon Bowers, *Healthy Native Stocks of Anadromous Salmonids in the Pacific Northwest and California*, prepared for Oregon Trout, Portland, Ore., December 31, 1994.

112. Reports and discussions of conflicts from PRC, op. cit. note 79; Alkire, op. cit. note 106; Paul Koberstein, "The Decline and Fall of Salmon," *High Country News*, November 15, 1993; Mark Clayton, "Latest Fish Fight: 'Captain Canada' Takes on Alaska," *Christian Science Monitor*, July 12, 1995; Bernard Simon, "Canada Closes Sockeye Salmon Fishery," *Financial Times*, August 12, 1995; William DiBenedetto, "US, Canada Seek Salmon Mediator as Talks Go Belly Up," *Journal of Commerce*, August 3, 1995; Bob Holmes, "Fishermen and Loggers Square Up Over Salmon," *New Scientist*, April 29, 1995; Bob Holmes, "Saving Snake River's Wild Salmon," *New Scientist*, April 22, 1995; Charles McCoy, "Regulators Slash Salmon Talk in West, Highlighting Threat to Fish's Survival," *Wall Street Journal*, April 13, 1992; tribal issues from Sterne, op. cit. note 79; Timothy Egan, "Indians of Puget Sound Get Rights to Shellfish," *New York Times*, January 27, 1995; Columbia River Inter-Tribal Fish Commission (CRITFC), *WY-KAN-USH-MI WA-KISH-WIT: Spirit of the Salmon* (Portland, Ore.: 1995); NRC, op. cit. note 79.

113. NRC, op. cit. note 79; Diamant and Wiley, op. cit. note 107; Karen Garrison and David Marcus, *Changing the Current: Affordable Strategies for Salmon Restoration in the Columbia River Basin* (San Francisco, Calif.: Natural Resources Defense Council, December 1994); Gunderson et al., eds., op. cit. note 65; Christopher A. Frissell and David Bayles, "Ecosystem Management and the Conservation of Aquatic Biodiversity and Ecological Integrity," *Water Resources Bulletin*, April 1996 (in press); CRITFC, op. cit. note 112; Frissell, op. cit. note 110; Alkire, op. cit. note 106; Wilderness Society, op. cit. note 110; Lichatowich and Mobrand, op. cit. note 111; Huntington, Nehlsen, and Bowers, op. cit. note 111; Henjum et al., op. cit. note 110; PRC, op. cit. note 79; Sterne, op. cit. note 79.

114. **Table 3** from Moyle and Leidy, op. cit. note 5; except Asia from Brian Groombridge, ed., *Global Biodiversity: Status of the Earth's Living Resources* (New York: Chapman and Hall, 1992); North America extinct and imperiled from The Nature Conservancy 1996, op. cit. note 3; Mexico from Salvador Contreras-B. and M. Lourdes Lozano-V., "Water, Endangered Fishes, and Development Perspectives in Arid Lands of Mexico," *Conservation Biology*, June 1994; Lake Victoria from Kaufman, op. cit. note 95 and Lowe-McConnell, op. cit. note 94.

115. David H.L. Thomas, "Artisenal Fishing and Environmental Change in a Nigerian Floodplain Wetland," *Environmental Conservation* Summer 1995; economic estimates in Pearce and Moran, op. cit. note 8.

116. Pearce and Moran, op. cit. note 8.

117. FAO op. cit. note 71.

118. Goodland, op. cit. note 37.

119. Karl-Geert Malle, "Cleaning up the River Rhine," *Scientific American*, January 1996.

120. Federal Interagency Floodplain Management Task Force, *A Unified National Program for Floodplain Management 1994* (Washington, D.C.: U.S. Federal Emergency Management Agency, 1995).

121. National Research Council, op. cit. note 79.

122. John B. Loomis, "Measuring the Economic Benefits of Removing Dams and Restoring the Elwha River: Results of a Contingent Valuation Survey," *Water Resources* (in press).

123. National Research Council, op. cit. note 27; Mississippi/Missouri restoration from Tim Searchinger, Environmental Defence Fund, personal communication with the author, November 22, 1995.

124. Paul L. Angermeier and James R. Karr, "Biological Integrity versus Biological Diversity as Policy Directives," *Bioscience*, November 1994.

PUBLICATION ORDER FORM

No. of
Copies

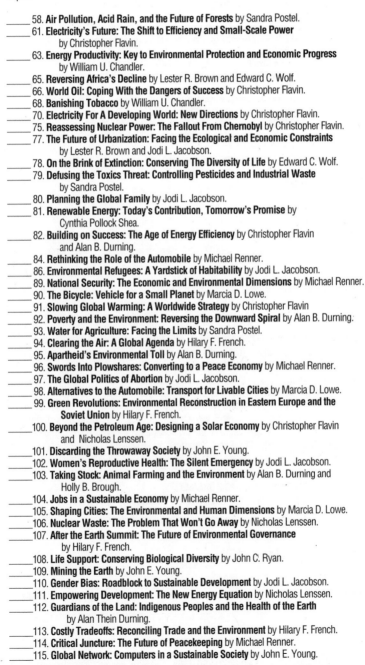

_____ 58. **Air Pollution, Acid Rain, and the Future of Forests** by Sandra Postel.
_____ 61. **Electricity's Future: The Shift to Efficiency and Small-Scale Power**
            by Christopher Flavin.
_____ 63. **Energy Productivity: Key to Environmental Protection and Economic Progress**
            by William U. Chandler.
_____ 65. **Reversing Africa's Decline** by Lester R. Brown and Edward C. Wolf.
_____ 66. **World Oil: Coping With the Dangers of Success** by Christopher Flavin.
_____ 68. **Banishing Tobacco** by William U. Chandler.
_____ 70. **Electricity For A Developing World: New Directions** by Christopher Flavin.
_____ 75. **Reassessing Nuclear Power: The Fallout From Chernobyl** by Christopher Flavin.
_____ 77. **The Future of Urbanization: Facing the Ecological and Economic Constraints**
            by Lester R. Brown and Jodi L. Jacobson.
_____ 78. **On the Brink of Extinction: Conserving The Diversity of Life** by Edward C. Wolf.
_____ 79. **Defusing the Toxics Threat: Controlling Pesticides and Industrial Waste**
            by Sandra Postel.
_____ 80. **Planning the Global Family** by Jodi L. Jacobson.
_____ 81. **Renewable Energy: Today's Contribution, Tomorrow's Promise** by
            Cynthia Pollock Shea.
_____ 82. **Building on Success: The Age of Energy Efficiency** by Christopher Flavin
            and Alan B. Durning.
_____ 84. **Rethinking the Role of the Automobile** by Michael Renner.
_____ 86. **Environmental Refugees: A Yardstick of Habitability** by Jodi L. Jacobson.
_____ 89. **National Security: The Economic and Environmental Dimensions** by Michael Renner.
_____ 90. **The Bicycle: Vehicle for a Small Planet** by Marcia D. Lowe.
_____ 91. **Slowing Global Warming: A Worldwide Strategy** by Christopher Flavin
_____ 92. **Poverty and the Environment: Reversing the Downward Spiral** by Alan B. Durning.
_____ 93. **Water for Agriculture: Facing the Limits** by Sandra Postel.
_____ 94. **Clearing the Air: A Global Agenda** by Hilary F. French.
_____ 95. **Apartheid's Environmental Toll** by Alan B. Durning.
_____ 96. **Swords Into Plowshares: Converting to a Peace Economy** by Michael Renner.
_____ 97. **The Global Politics of Abortion** by Jodi L. Jacobson.
_____ 98. **Alternatives to the Automobile: Transport for Livable Cities** by Marcia D. Lowe.
_____ 99. **Green Revolutions: Environmental Reconstruction in Eastern Europe and the**
            Soviet Union by Hilary F. French.
_____ 100. **Beyond the Petroleum Age: Designing a Solar Economy** by Christopher Flavin
            and Nicholas Lenssen.
_____ 101. **Discarding the Throwaway Society** by John E. Young.
_____ 102. **Women's Reproductive Health: The Silent Emergency** by Jodi L. Jacobson.
_____ 103. **Taking Stock: Animal Farming and the Environment** by Alan B. Durning and
            Holly B. Brough.
_____ 104. **Jobs in a Sustainable Economy** by Michael Renner.
_____ 105. **Shaping Cities: The Environmental and Human Dimensions** by Marcia D. Lowe.
_____ 106. **Nuclear Waste: The Problem That Won't Go Away** by Nicholas Lenssen.
_____ 107. **After the Earth Summit: The Future of Environmental Governance**
            by Hilary F. French.
_____ 108. **Life Support: Conserving Biological Diversity** by John C. Ryan.
_____ 109. **Mining the Earth** by John E. Young.
_____ 110. **Gender Bias: Roadblock to Sustainable Development** by Jodi L. Jacobson.
_____ 111. **Empowering Development: The New Energy Equation** by Nicholas Lenssen.
_____ 112. **Guardians of the Land: Indigenous Peoples and the Health of the Earth**
            by Alan Thein Durning.
_____ 113. **Costly Tradeoffs: Reconciling Trade and the Environment** by Hilary F. French.
_____ 114. **Critical Juncture: The Future of Peacekeeping** by Michael Renner.
_____ 115. **Global Network: Computers in a Sustainable Society** by John E. Young.

_____ **Total Copies**

Single Copy: $5.00   • 2–5: $4.00 ea.   •   6–20: $3.00 ea. • 21 or more: $2.00 ea.
Call Director of Communication at (202) 452-1999 to inquire about discounts on larger orders.

☐ **Membership in the Worldwatch Library: $30.00 (international airmail $45.00)**
The paperback edition of our 250-page "annual physical of the planet,"
State of the World, plus all Worldwatch Papers released during the calendar year.

☐ **Subscription to *World Watch* magazine: $20.00 (international airmail $35.00)**
Stay abreast of global environmental trends and issues with our award-winning, eminently readable bimonthly magazine.

☐ **Worldwatch Database Disk Subscription: One year for $89**
Includes current global agricultural, energy, economic, environmental, social, and military indicators from all current Worldwatch publications. Includes a mid-year update, and *Vital Signs* and *State of the World* as they are published. Can be used with Lotus 1-2-3, Quattro Pro, Excel, SuperCalc and many other spreadsheets.
Check one: _____high-density IBM-compatible or _____Macintosh

Make check payable to Worldwatch Institute
1776 Massachusetts Avenue, N.W., Washington, D.C. 20036-1904 USA

Please include $3 postage and handling for non-subscription orders.

Enclosed is my check for U.S. $_____
AMEX ☐ VISA ☐ Mastercard ☐ _____

Card Number   Expiration Date

name   **daytime phone #**

address

city   **state   zip/country**

Phone: (202) 452-1999   Fax: (202) 296-7365   E-Mail: wwpub@igc.apc.org

WWP

PUBLICATION ORDER FORM

No. of
Copies

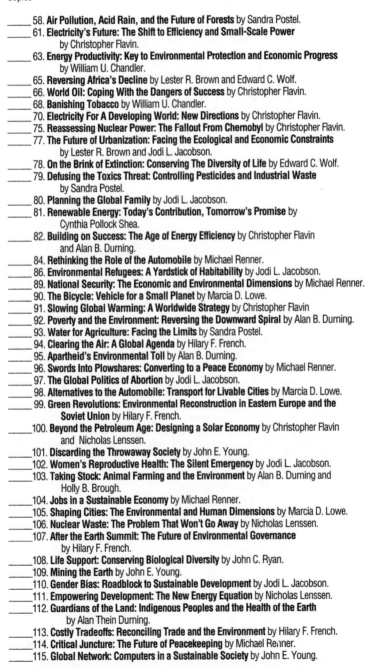

_____ 58. **Air Pollution, Acid Rain, and the Future of Forests** by Sandra Postel.

_____ 61. **Electricity's Future: The Shift to Efficiency and Small-Scale Power** by Christopher Flavin.

_____ 63. **Energy Productivity: Key to Environmental Protection and Economic Progress** by William U. Chandler.

_____ 65. **Reversing Africa's Decline** by Lester R. Brown and Edward C. Wolf.

_____ 66. **World Oil: Coping With the Dangers of Success** by Christopher Flavin.

_____ 68. **Banishing Tobacco** by William U. Chandler.

_____ 70. **Electricity For A Developing World: New Directions** by Christopher Flavin.

_____ 75. **Reassessing Nuclear Power: The Fallout From Chernobyl** by Christopher Flavin.

_____ 77. **The Future of Urbanization: Facing the Ecological and Economic Constraints** by Lester R. Brown and Jodi L. Jacobson.

_____ 78. **On the Brink of Extinction: Conserving The Diversity of Life** by Edward C. Wolf.

_____ 79. **Defusing the Toxics Threat: Controlling Pesticides and Industrial Waste** by Sandra Postel.

_____ 80. **Planning the Global Family** by Jodi L. Jacobson.

_____ 81. **Renewable Energy: Today's Contribution, Tomorrow's Promise** by Cynthia Pollock Shea.

_____ 82. **Building on Success: The Age of Energy Efficiency** by Christopher Flavin and Alan B. Durning.

_____ 84. **Rethinking the Role of the Automobile** by Michael Renner.

_____ 86. **Environmental Refugees: A Yardstick of Habitability** by Jodi L. Jacobson.

_____ 89. **National Security: The Economic and Environmental Dimensions** by Michael Renner.

_____ 90. **The Bicycle: Vehicle for a Small Planet** by Marcia D. Lowe.

_____ 91. **Slowing Global Warming: A Worldwide Strategy** by Christopher Flavin

_____ 92. **Poverty and the Environment: Reversing the Downward Spiral** by Alan B. Durning.

_____ 93. **Water for Agriculture: Facing the Limits** by Sandra Postel.

_____ 94. **Clearing the Air: A Global Agenda** by Hilary F. French.

_____ 95. **Apartheid's Environmental Toll** by Alan B. Durning.

_____ 96. **Swords Into Plowshares: Converting to a Peace Economy** by Michael Renner.

_____ 97. **The Global Politics of Abortion** by Jodi L. Jacobson.

_____ 98. **Alternatives to the Automobile: Transport for Livable Cities** by Marcia D. Lowe.

_____ 99. **Green Revolutions: Environmental Reconstruction in Eastern Europe and the Soviet Union** by Hilary F. French.

_____100. **Beyond the Petroleum Age: Designing a Solar Economy** by Christopher Flavin and Nicholas Lenssen.

_____101. **Discarding the Throwaway Society** by John E. Young.

_____102. **Women's Reproductive Health: The Silent Emergency** by Jodi L. Jacobson.

_____103. **Taking Stock: Animal Farming and the Environment** by Alan B. Durning and Holly B. Brough.

_____104. **Jobs in a Sustainable Economy** by Michael Renner.

_____105. **Shaping Cities: The Environmental and Human Dimensions** by Marcia D. Lowe.

_____106. **Nuclear Waste: The Problem That Won't Go Away** by Nicholas Lenssen.

_____107. **After the Earth Summit: The Future of Environmental Governance** by Hilary F. French.

_____108. **Life Support: Conserving Biological Diversity** by John C. Ryan.

_____109. **Mining the Earth** by John E. Young.

_____110. **Gender Bias: Roadblock to Sustainable Development** by Jodi L. Jacobson.

_____111. **Empowering Development: The New Energy Equation** by Nicholas Lenssen.

_____112. **Guardians of the Land: Indigenous Peoples and the Health of the Earth** by Alan Thein Durning.

_____113. **Costly Tradeoffs: Reconciling Trade and the Environment** by Hilary F. French.

_____114. **Critical Juncture: The Future of Peacekeeping** by Michael Renner.

_____115. **Global Network: Computers in a Sustainable Society** by John E. Young.

_____ **Total Copies**

Single Copy: $5.00 • 2–5: $4.00 ea. • 6–20: $3.00 ea. • 21 or more: $2.00 ea.
Call Director of Communication at (202) 452-1999 to inquire about discounts on larger orders.

☐ **Membership in the Worldwatch Library: $30.00 (international airmail $45.00)**
The paperback edition of our 250-page "annual physical of the planet,"
State of the World, plus all Worldwatch Papers released during the calendar year.

☐ **Subscription to *World Watch* magazine: $20.00 (international airmail $35.00)**
Stay abreast of global environmental trends and issues with our award-winning,
eminently readable bimonthly magazine.

☐ **Worldwatch Database Disk Subscription: One year for $89**
Includes current global agricultural, energy, economic, environmental, social, and
military indicators from all current Worldwatch publications. Includes a mid-year
update, and *Vital Signs* and *State of the World* as they are published. Can be used
with Lotus 1-2-3, Quattro Pro, Excel, SuperCalc and many other spreadsheets.
Check one: _____high-density IBM-compatible or _____Macintosh

---

Make check payable to Worldwatch Institute
1776 Massachusetts Avenue, N.W., Washington, D.C. 20036-1904 USA

Please include $3 postage and handling for non-subscription orders.

Enclosed is my check for U.S. $_____
AMEX ☐ VISA ☐ Mastercard ☐ _____
Card Number   Expiration Date

name   **daytime phone #**

address

city   **state   zip/country**
Phone: (202) 452-1999   Fax: (202) 296-7365   E-Mail: wwpub@igc.apc.org   WWP